DRAGONS IN THE SNOW

EDWARD POWER

DRAGONS IN THE SNOW

Avalanche Detectives and the Race to Beat Death in the Mountains

MOUNTAINEERS BOOKS

MOUNTAINEERS BOOKS is dedicated
to the exploration, preservation, and enjoyment
of outdoor and wilderness areas.

1001 SW Klickitat Way, Suite 201, Seattle, WA 98134
800-553-4453, www.mountaineersbooks.org

Published simultaneously in Great Britain and North America in 2020 by Vertebrate Publishing, Sheffield, and Mountaineers Books, Seattle.

Printed in the United States of America
23 22 21 20 1 2 3 4 5

Copyeditor: Emma Lockley
Cover design: Nathan Ryder, www.ryderdesign.studio
Layout: Jen Grable
Cover photograph: *Ethereal early morning light on summit day, February 2, 2011. Denis Urubko, Simone Moro and Cory Richards became the first to climb Gasherbrum II in winter.* © Cory Richards
Frontispiece: *Simone Moro on the summit of Makalu, February 9, 2009, when he and Denis Urubko made the first winter ascent of the mountain.* © Denis Urubko

Library of Congress Control Number: 2020938923

Mountaineers Books titles may be purchased for corporate, educational, or other promotional sales, and our authors are available for a wide range of events. For information on special discounts or booking an author, contact our customer service at 800-553-4453 or mbooks@mountaineersbooks.org.

Printed on FSC-certified materials

ISBN (paperback): 978-1-68051-292-2
ISBN (ebook): 978-1-68051-293-9

An independent nonprofit publisher since 1960

For Marguerite, who is far more
than my ski partner

I'll tell you what's interesting about snow. Ninety-nine times out of a hundred, nature allows you to get away with it. Then there is one time, maybe, that you don't recognize something is slightly different. And I think that's probably when most experienced people get tricked.

—Craig Gordon, forecaster, Utah Avalanche Center

Reading snow is like listening to music. To describe what you've read is like explaining music in writing.

—Peter Hoeg, *Smilla's Sense of Snow*

Contents

Author's Note

When I moved full-time to Northern Utah in January 2016, I brought with me a thirty-year career of curiosity as a newspaper journalist, and decades of enthusiasm for recreational skiing. In those first months, I was struck by the almost daily, early morning booms of avalanche mitigation bombs echoing across the valley in front of my home. As an amateur skier, I was inspired to learn more about the avalanche and backcountry skiing professionals who do this work. The result is this book based on extensive in-person or, in some cases, phone interviews with the people whose stories are recounted here. Documents related to avalanche advisories, accident investigations, written survivor accounts, weather data, websites, and other sources were also heavily utilized, and I have sought to clearly attribute those sources. With few exceptions, interviews were recorded and transcribed for narrative accuracy. Many of the people quoted in this book reviewed drafts of chapters in which their stories and voices appear. Material from newspapers, magazines, and books is duly attributed.

The risks in backcountry skiing and snowboarding are part of the allure as the rider ventures into wilderness terrain and often pristine snow. But anyone undertaking these activities should do so only after receiving avalanche awareness training and obtaining such specialized gear as a beacon, probe pole, and shovel. Traveling with a skilled partner is essential.

Emergency winter rescues in the mountains are, by nature, extremely dangerous. They are fast-paced, may involve makeshift medical procedures, often involve moments of confusion, and run on high human emotions. I have strived to accurately represent the heroic work of snow professionals and nonprofessionals whose selfless efforts have, without question, saved many lives. Any misrepresentations of facts, attributions, or other details related to these stories are unintended.

Prologue: Tail of the Dragon

On January 5, 2017, nearly two hundred miles off the coast of northwestern North America, above the seething, frigid waters of the Gulf of Alaska, an evolutionary moment began. It was unseen by human eyes. Even the roving, photographic gaze of a geostationary weather satellite, orbiting 22,300 miles above the equator, did not at first distinguish this cold whiff of wind and water. Many months later, a weather scientist named Randy Graham, working out of an office in Salt Lake City, Utah, would describe it as a bare "finger," like a tiny fetus forming. Once it had grown to far larger proportions, he retroactively altered his categorization, dubbing it a "fire hose."

But it was really more like a dragon. Uncoiling from the milky, numbing waters of the Alaskan gulf, its birth was a dragon's tail dipping into the waves of the notoriously anarchic gulf, feeding the buildup of its body and head. For Graham and his colleagues at the National Oceanic and Atmospheric Administration (NOAA) offices, the tail grew into a white smear of water vapor across hundreds of square miles of the North Pacific. They saw it grow in time lapse, fifteen-minute increments, through daytime photos, and into nighttime infrared images, blue-hued water clouds swaddling the dragon's back, and white ice clouds rimming its sinuous jaws.

The storm was still more of a meteorological serpent than what avalanche forecasters would later call a dragon in the snow. But it grew quickly. The flourishing dragon acquired motion, its tongue of wind sweeping more moisture up with its elongating tail. As that tail dipped deeper into the Pacific, it grew larger, whipping the entire length of its mass into a spinning frenzy of water and warm, tropical wind sucked from the south. The storm crawled aggressively across the ocean's surface, stirring ever bigger waves like deep, maritime claw marks.

In the air, seabirds raced to escape it. Beneath the roiling sea surface, fish sensed the disintegrating world above. Though still far out to sea, this storm dragon, becoming a yet unseen creature from ashore, began to take on more form. From space, it was a white blur of inchoate energy, yet it was learning form, strengthening, gathering a more precise direction. Within a few days, the tail of the storm dragon had become massive, sucking more and more energy into its fattening belly. Instinct seemed to have taken over, and the maw of it inarguably pointed southeast.

Had forecasters studied it from the beginning, they could have charted its immature yet undeniable track. It was headed for Oregon and then into the Sierra Nevada of Northern California. Soon, under the darkness of night, this snow dragon was hurtling toward the coastline, making landfall. In the hours ahead, it was revealed as a monster fully born.

Once it had crossed nearly two hundred miles of the California mainland, the storm initiated its climb into the Sierra on January 9. So thick and heavy were its clouds and snow, it obliterated light and the expanse of daytime sky. The usual horizon of cross-stitched mountain peaks disappeared. Anything that might have suggested a redoubt of human survival seemed to have vanished. Snow, usually so welcome in California's parched terrain, turned to something more like the delivery of a plague.

While the snow dragon was its own beast, it was the fourth such storm in a matter of seven days. More than twenty feet of snow lay like a new, frozen ocean over the mountaintops. Homes and cars were drowned in it. In a place where it would have seemed to be most welcome—Mammoth Mountain ski area—the storm had disabled ski lifts, its snow so deep that lift chairs could not ferry skiers. After forty-eight seasons of skiing, Mammoth recorded the most snowfall in a single month—246 inches, more than 20 feet. The ski resort's historic average for January had been 65 inches. Even more imposing, the January snows alone that year had already topped eleven previous ski season totals.

As the storm rampaged its way over the Sierra, Interstate 80 through the infamous Donner Pass was shut down. People referred to the month as "Janburied." But it was this specific storm dragon from the Gulf of Alaska that reigned supreme over the Sierra. Once it had climbed into the freezing mountain air on January 9, it jettisoned seventy-seven inches of snow across Sugar Bowl ski area; in a forty-eight-hour period at Kirkwood ski area, seventy-one inches fell. In the small settlements of Kingvale and Soda Springs, barely five miles west of Donner Pass, residents saw seventy-nine inches—six and a half feet—fall in two days.

As it crested like a freezing tsunami above the Sierra, the snow dragon left behind stands of evergreens snapped like vertebrae, stranded cars and eighteen-wheelers, severed power lines, and darkened, empty mountain town streets. Then it fixed its leading edge, its snow-spitting jaws, squarely to the east. Spread out in its path was a new mountainous quarry—the already snow-packed summits of Utah's Wasatch and Uinta ranges.

1

ON A HIGH SLOPE IN THE UINTAS

In the shadow of Utah's Uinta Mountains, early on the morning of January 11, 2017, Jeremy Jones, a forty-one-year-old pro snowboarder from Utah, stands at dawn in a parking lot in Oakley, a tiny, blink of a town whose main feature is an aluminum-clad eatery, a throwback to the 1950s called the Road Island Diner. But Jones isn't scoping out a breakfast spot on this Wednesday morning. He's assessing a near blizzard of a snowstorm that has been cresting over the mountains for nearly twelve hours. Standing there in front of the dark windows of Dutch's service station, Jones is overseeing the rendezvous of a group of backcountry snowboarders planning to make an assault on a vast face of terrain overlooking the Smith-Morehouse Reservoir.

The snowstorm that Jones is surveying has marched into Utah's northern mountain ranges after a long weekend of two other strong fronts that left chest-deep powder across the ridgelines and valleys. Nearly five feet has fallen. Jones marvels at it, and at the group of snowboarders gathering before him. Even with the heavy snow's insulation nearly silencing all small talk, anticipation crackles through the searing morning air, as if the dense grid of snowflakes are radio crystals emitting urgent public weather alerts in the light wind.

Jones herds the snowboarders together. Including himself, there are nine riders. The plan is to proceed from Oakley to the Smith-Morehouse trailhead, where they will meet up with a snowcat—a vehicle part snowplow, part tank, capable of ferrying them through deep snow into the backcountry. Jones has pulled together this trip for friends and professional acquaintances to, as he puts it, "show 'em a good day."

Saddling up in their cars, the group arrives at the trailhead as the snowcat is unloaded from its trailer. Soon, its powerful diesel engine fires up and the cat hums as it waits, expressing its own kind of cautious, mechanized suspense about what the day will bring. It's hard for Jones to keep the hardcore boarders' attention off the cat, which they equate with a magic carpet ride into the backcountry. But Jones needs the riders, some of whom don't know each other well, to listen to his plan before the trail leads them deep into the mountains.

The storm is "nuking," as Jones characterizes it, shedding so much snow that it is likely to cover the snowcat's tracks as it winds its way into the Uintas. While this doesn't raise anyone's concern about their return on the same trail, it emphasizes the sheer force of what the riders are heading into. This is no day for backcountry novices.

FOR GOING ON TWENTY-FIVE YEARS, Jones, not to be confused with the famed snowboarder and conservationist of the same name based in Truckee, California, has built a mindset, one could even say a methodology, around staying alive in the winter cold and hip-deep snow of Utah's remote backcountry. It's what some riders call the out-of-bounds, the antithesis to in-bounds, or resort skiing and snowboarding. Adventure and untracked snow are the lures, but it is more than that. Backcountry skiers, snowboarders, and even snowmobilers, feel that when they are removed from the oversight of ski patrollers and others, they are immersed in a world of self-reliance, breaking their own trails, free to cut heroic lines down mountain faces while dodging injury and death.

The survival gear Jones and three friends have fashioned over the years is packed away in both his mental storage pocket and a literal one in his pack. He prepares for the out-of-bounds by having "risk talks" with himself, going through the litany of avalanche threat signposts: snow cornices that can crash down like ocean waves and sweep a body away; telltale cracks in the snowpack that mark avalanches already brimming under the surface; trees where branches have been ripped away by avalanche activity. Within the physical pouch of tools Jones carries—beyond his own avalanche beacon, shovel, and snow probe pole—is a simple survival kit: a knife, a good length of cordage, some basic first-aid supplies, and something to ignite a fire. Jones lives by a personal creed about backcountry risks: "If you're not prepared to fucking survive—you won't."

Having engaged in his own private risk talk that morning, he turns to some of the snowboarders who have checked the Utah Avalanche Center's advisory. One particular report has caught their attention; it's a forecast warning of "extreme" danger in the Uintas, a rare black-shaded classification that sits at the top of a five-tier scale of warnings that begin with green-shaded "low."

"It had been seen by multiple people," Jones says. "We knew that going in. We had decided we would stay in thick trees." Under the threat of avalanches, it's a common tactic, the snowboarders know, to make runs in glades of trees rather than across steep, open mountain faces. Snow is usually more stable in thick trees, for several reasons. Trees tend to grow on lower-angled terrain and their trunks help to anchor the snow. There is less chance of wind loading up snow pillows and other features where the touch of a human foot can trigger a slide.

Boarding the snowcat, the riders start to claim seating and stash their gear. All their snowboards have been secured in metal baskets affixed to the cat's exterior metal shell. The snowboarders settle in on the two bench seats that run the length of the passenger cabin. Backpacks and other gear are stowed beneath the benches. A few things, including gloves and goggles, are tossed into small, hammock-like nets dangling from the ceiling. Boots bang on the steel floor, but the noise fades as the riders settle into their private excitement or anxiety.

The snowcat's two drivers wedge into their seats and throttle up the engine. The big cat, with its shrugging, elephantine gait, shuffles to the trailhead. The way ahead is covered in a tidal wave of snow. As Jones takes his place on a bench seat, he hears the drivers' music. The tunes are emo—too slow for Jones's tastes. But as he takes in the morning scene and the body language of the riders, he realizes it may be just as well. "Everyone was a little more somber than on a usual powder day," he says. "We were more cooled down, not as jumpy or amped."

Maybe it's the lingering shadow of the avalanche advisory, the overwhelming thrust of snow falling outside the cat's windows, or even the unfamiliarity of some of the riders with each other. Jones senses a subdued nervousness among the group, so he's happy when the snowcat arrives at a clearing that includes some toilet facilities about two miles from where they left their cars. It's about 8:30 a.m. and a hint of daylight seeps through the storm clouds. He can get everyone outside, get them moving again, focused on becoming a more close-knit group.

Each of the riders has brought along the requisite avalanche beacon, a device strapped to the chest which emits a silent "beeping" signal that can be tracked by other beacons in "search" mode. Digital numbers on "search" beacons display the distance to beacons set in "send" worn by a skier or boarder trapped under avalanche debris. A rule of thumb for avalanche rescue is that a survival clock starts to tick as soon as someone is buried; at about ten minutes the chances of digging someone out alive decrease like water in a tub making its final spin down a drain. Jones takes beacon practice seriously. He's been caught in four slides in his snowboarding career, each time fortunate to remain above the snow. He understands the sheer, sudden acceleration and power of an avalanche.

The snowstorm that morning has been nearly blinding at times. But it suddenly backs off over the clearing as if welcoming the human presence and doing its part to facilitate beacon rescue practice. A cold fog hangs above the terrain like an expression of the snowboarders' mood. Jones perceives the storm's character, since the previous night has been one of intermittent whiteouts and brief lapses in snowfall that some skiers call "sucker holes"—false promises of an abating storm, particularly when the sun shines through for a few merciful minutes.

With a lot of snowcat travel still ahead of them, Jones instructs a couple of the riders to bury their beacons, switched to "send" mode, in the deep snow of the clearing. He shouts to the remaining boarders to switch their beacons to "search." Quickly they start hunting down the signals, following readouts and lines and arrows on their beacon screens indicating distance and direction.

Jones is attentive as he watches the beacon practice. He wants to make sure everyone's equipment is functioning, and he wants to gauge the expertise and demeanor each rider shows in an avalanche rescue simulation. While the search is only practice, the gray light and heavy snow of the morning lend it an air of the real thing. Tentativeness or confusion in practice, Jones knows, does not bode well for someone's reactions deep in the backcountry. When real people are buried many feet under the snow, panic and confusion can replace the calm of a simulation. With no signs of the person's location, no ski pole protruding from the snow, no ski or snowboard visible in the cement-like morass of a big slide, searchers easily lose focus. They get caught up in the desperate minutes that feel like final sand particles in a tiny hourglass.

In many ways the makeup of the group personifies the growing tension between the backcountry and the young practitioners of the skills needed in the wild. Some commit to the art and science of it—climbing skins, skis or snowboards of a certain design, route finding, careful dissection of snow layers. Others simply want the adrenaline charge of leaving behind civilization and strapping in to ride the thin edge of danger for a day. All this is placing more pressure on the virgin terrain and on the professionals responsible for oversight—avalanche forecasters, rescue crews, medical operations.

Jones is looking to identify the novice out-of-bounds riders. After thirty or forty minutes, he decides everyone has bonded a bit and understands the basic plan and techniques for a victim search. With a final verbal check of each rider—"Are you beeping?"—Jones and one of his partners wave their "search" mode beacons across the riders' chests and receive reassuring confirmation that the beacons are broadcasting.

Back aboard the snowcat, riders slide into their spots. As the diesel winds up and the cat's beastlike machinery resumes lumbering higher on the trail to the peaks above, Jones takes stock of the cabin's mood. He's sensitive to the earlier unease, and although it isn't completely erased by the beacon practice, the air seems to contain a healthier vibe. Even if some anxiety remains—something that might be interpreted as a negative, a waning confidence felt by a few riders—it doesn't strike Jones as a contaminant to the group's mindset. In some ways he takes this read as a good sign. The crew has been sobered by the practice and its echo of the morning's avalanche warning.

Jones intuits that in the exercise a kind of pecking order has been established; more experienced backcountry riders have naturally claimed leadership roles should there become a need for swift action. Surveying the group, he fixes on his core rider friends, those he's practiced with and headed into the backcountry with for more than two decades: Mike Nelson, Seth Huot, and Brock Harris. There's a fourth member of the brotherhood, J. P. Walker, but he's not on board. Having three of his buddies with him this morning reassures Jones. When assembling the group, he was sure to include "people I trusted," his "foxhole buddies," convinced that in whatever battle they might encounter—against man or nature—they would prevail.

This is no small claim for Jones. Looking back to his earliest days of snowboarding, the road is long and traveled with a purpose and commitment Jones is proud of. He has had to think on his feet as the snowboard industry has evolved and money for pro riders has dried up or changed dramatically.

Growing up in Farmington, Utah, Jones was a skateboarder, with limited experience on the snow. But around age twelve, he saw the emerging presence of snowboarding in the slopes near home. He was convinced it was a sport that his skateboarding prowess could be adapted to so he pried the wheels off of one of his skateboards and then cut two lengths of bicycle tire tubing from which he fashioned crude bindings that he fastened to the deck for his feet.

In friends' backyards during winter, Jones practiced moves on his makeshift snowboard, tricks translated from his skateboarder's repertoire. After two winters of refining his agility on snow, Jones, now fourteen, finally got a real snowboard and began to ride in resorts, particularly Brighton ski area in Big Cottonwood Canyon, south of Salt Lake City. There, Jones and his core group bonded, founding a kind of private, insular backcountry enclave they christened "The Spot," within which they built their skillset for deeper, out-of-bounds exploration. They developed their small rescue toolkit of knife, rope, first aid, and fire starter. They schooled themselves in identifying signs of potential avalanches. And they worked to create among themselves a hand-in-glove response to a slide, a mental rescue template buttressed by cool heads and advanced athleticism.

Of those early winters and his buddies at Brighton, Jones says, it was a process of convening to "tune up our minds, tune up our bodies." As they spread their out-of-bounds riding into steeper, more remote terrain, they took to filming their exploits. It was the early days of such pursuits, and they carved out a distribution deal for their snowboard films with a company called Mack Dog Productions. The formula was simple: round up twelve to fifteen riders, give each one three minutes or so of explosive exposure, and voilà: a new release was on tap. Priced at ten bucks or so, the films would sell ten thousand copies, and Jones and his team would make enough money to fund their winter riding. This went on for years—until the proliferation of similar short films on YouTube and other media essentially glutted the market.

More recently, Jones and his friend Mike Nelson decided to expand their footprint in the snowboarding world. They founded a small company named Destroyer to produce protective gear for riders, including helmets, padding, and a small line of soft goods. For Jones it's not only a new entrepreneurial path but also an acknowledgment that at forty-one he has a wife, Sher, and two children, Adi, his twelve-year-old daughter, and Cru, his ten-year-old son.

IN THE CRAMPED CONFINES OF the snowcat, Jones thinks of his family then directs his attention to the setting before him. The conditions inside the cat aren't as luxurious as those in some ski resort cats where the seats are thickly padded, there's ample heat, and chests brimming with hot soup, sandwiches, snacks, and endless bottles of cold water to offset dehydration from altitude. As the RPMs increase and the machine climbs the steeper ridges higher into the Uintas, a little heat from the engine wafts back into the cabin. The riders are well equipped, experienced in winter conditions, and most have brought packs with extra gear.

As if the ponderous pace of the snowcat isn't slow enough, Jones sees that they've encountered a new obstacle—a tree fallen across the trail. To try to maneuver the vehicle over it would risk damage to the snow treads or undercarriage of the machine. The only way to clear the path is to dispatch a couple of riders with saws to cut the tree into pieces. Luckily, a couple of saws are on board, typically used for slicing into snowpack so that snow pits can be assessed for layers of weak or "sugary" snow that is avalanche prone.

The saw party is quick with their work and the snowcat powers up again. They have been going for nearly three and a half hours. It's early afternoon and they still haven't made a single run. Tension has slipped back into the cabin, and Jones is growing edgy to get into the deep powder he and the others see accumulating. It's a snowboard Shangri-la out there, but they have yet to experience any runs in it.

Finally, they arrive at a spot, not at the mountain's ultimate peak but adjacent to a broad stand of trees that fits the description of what they agree is likely safe terrain. It's a lower-angle slope—about twenty-seven to twenty-nine degrees in pitch—significantly under the high thirties or even forty-degree slope terrain they might tackle in less threatening conditions. Outside the snowcat, they scan the glade of pines below them as the snow continues to cascade down. "It was almost a violent storm," Jones says. "You could get out and you could almost feel the anxiety of the storm but it was also calming." The contrasting moods of the storm leave Jones feeling isolated from his normal "read" of such a setting, its slopes an ambivalent vista before him. He's wary of the day's avalanche threat, yet there is serenity all around him—puffs of snowy ornamentation in the tree branches, softer-slope angles, a quiet peace reminiscent of a Christmas snow globe.

"Everyone is anxious to get on the snow," he says. "We're certain it's not going to slide." With the riders geared up and ready to launch into the lines before them, Jones packs a walkie-talkie into his jacket and declares he will make a test run down through the trees. The eight riders see Jones drop in and thread his way through the trees, a white smoke of snow flying skyward in his wake. When Jones reaches a logical stopping point, he rounds up, pulls out his radio, and reports: "It's good. Follow my run!"

Watching the riders come down like acrobatic warriors, Jones is still in the grasp of his own ride. "It was some of the best snow ever in my career," he says. "It was insane. Everyone was spitting snow out of their mouths. I'm choking it out. I don't know who's who because their faces are butchered with snow." With the elation of their first run tingling in their legs and heads, the group decides to make a second run through the trees. It's another powder fest, an angel float of turn after turn as if they are colorful kites carving deep arcs in the wind. This is what they have come for, the day of perfection.

After dropping off the riders, the snowcat drivers motor to a lower position on the mountain where the boarders can reload their boards and slip into the cabin to hydrate, eat, or unwind for a few minutes. A welcome rhythm has set in. An editor and photographer from *Snowboarder* magazine are among the group to help document the day. There's an unspoken expectation that some riders may be featured in a spread.

In a time check, Jones sees it's almost 2:30 p.m. The last rider into the snowcat is his close friend Mike Nelson. It seems almost unnecessary, but Jones does a head count and confirms that everyone is comfortable with the conditions and the run. Only smiles and nods greet him.

"Let's go up higher and see what it looks like," Nelson says, and the snowcat steers for altitude. Compared to the earlier laborious travel, this stretch is quick, maybe ten minutes up a trail and onto a ridgeline, a true peak. It's a far different set-up than the gladed tree run. They are at about 9,400 feet, looking down a slope that's angled about thirty-seven degrees. It's the difference between having made an easy dive off a three-foot springboard and suddenly finding one's toes gripping the edge of a thirty-three-foot-high Olympic platform.

Just as with the tree runs, Jones reviews the plan for testing the slope. "We were all more cautious of this run," he says. From his perspective on the ridge, he can see the first long pitch drops down to a convex "roll" in the terrain before the slope disappears into a more vertical angle. He decides he will drop in and make a long, diagonal "ski cut" to the right, a maneuver skiers and boarders use to test the releasability of snow on precipitous terrain.

And then he's in. "I'm thinking it's going to go, get a little release," he says, "but it doesn't go. I hit it hard once, gave it two big hops over ten feet. I saw a little bit of a crack and some surface sluff. It gave me zero concern." At the end of the ski cut, Jones angles his snowboard down, collects speed, continues to accelerate, and makes a big turn back toward the center of the slope.

He's ripping down the face now, snow erupting off the tail of his board. He's purely in the boarder's paradise zone. But he's also drawing on all his veteran knowledge of the backcountry. One thing he has not forgotten to scope out—something all backcountry skiers and boarders are counseled to devise, even if only in a fraction of a second—is an exit route should an avalanche break. Jet to the far side, out of the slide's path, take refuge behind a tree or large rock. Whatever gets the rider out of the slide's flood of snow, ice, and debris before it overtakes them.

Racing toward the convex rollover below, Jones spots a big tree to the right side of the slope and executes another big maneuver with a bead on the tree. "I hacked a heavy turn, came around a pine tree and stopped. The turn felt so good." Inspecting the face for any evidence of sliding snow, Jones sees none. Dropping back into the slope, he rides about 150 feet farther down, stops on top of another rollover, and radios up to the other riders.

Down they come in clouds of snow light as feathers. Jones sees one boarder hit the first rollover and launch a jump he judges to be forty feet. Suspended in the air for what seems like a slow-motion film frame, the rider lands in a thick detonation of fluffy snow bursts as if shot from cannons. "We all got down another amazing run," Jones says.

Another time check: It's just before 4:00 p.m. Light is starting to drop, and Jones knows they face a three- to four-hour return trip to the trailhead. On almost any other day, he would lean toward starting back. The snow is unrelenting and, for all he knows, they could encounter another toppled tree blocking the path. But as Jones looks up toward the symmetry of tracks drawn by riders across the mountain face, he decides they should go for one last run.

Inside the snowcat, as the vehicle claws its way back up the ridgeline, Jones realizes this will be the final chance for the magazine photographer to claim some photos. He proposes positioning her to the left side of the slope, out of the direct path of the action but close enough to get clear shots. At the top of the ridgeline, having already tested the firmness of the snowpack, Jones directs another rider to ski cut across the slope to the right and stop. Jones takes a deep breath, eyes his imaginary line down the steepest part of the face, and leans to drop in.

"I hit the drop, I do a forward roll and then back to my feet," Jones says. The snow is so soft and light he can only smile at his antics. As he gains speed, passing the photographer off to his left, Jones locks his gaze on the track left by the snowboarder who veered to the right. Jones hits the ski cut, drops about five feet below it, and carves a hard turn toward the right side of the mountain. Then he hears it.

"The most violent thunder sound, like it hit right over the top of my house. . . . And then I see the ripples." Under Jones's feet, the earth is reforming itself. Something alive is bubbling up from under the snow surface. To his almost detached amazement, the white churn turns to a vortex around his knees, then his waist, clawing up his chest.

Faster and faster it goes until it feels to Jones like the entire face of the mountain has snapped free. And he is now part of it.

2

4:06 A.M. AVALANCHE ADVISORY

It is fully two hours past Craig Gordon's customary bedtime of 8:00 p.m. Still wide awake, he is coming down from the sensory rush of the address he delivered earlier this Tuesday night in Park City—a presentation at Treasure Mountain Middle School, where ski patrollers and others gathered to hear Gordon's analysis of the current 2016–17 winter snowpack. His talk is followed by a tense, white-knuckle drive home in a raging snowstorm that is shoveling itself onto the mountain ranges of northern Utah.

If the aftermath of Gordon's slideshow and fraught drive aren't troublesome enough, he feels an additional level of disquiet on this particular night in January. Not an unfamiliar anxiety to someone in his line of work. The faint but growing alarm Gordon senses has an urgency that is massing in his mind like the snow beyond the windows of his Cottonwood Heights home, just south of downtown Salt Lake City. A US Forest Service employee, Gordon works at the Utah Avalanche Center, where he is the sole forecaster for avalanche advisories in Utah's Uinta mountain range. As an avalanche awareness trainer, he has deep experience in backcountry skiing and rescue techniques.

The largest range in Utah, the Uintas are marked by a number of distinctions that make Gordon's job of avalanche forecasting even more dicey than one might imagine. It is the only major mountain range in the Lower 48 that runs west to east, which means the range receives weather systems far differently than those running north to south like the Wasatch. Low-pressure systems that flow from the Pacific toward the plains states and the Atlantic's Eastern Seaboard directly hit the slopes of the Wasatch, while the high-elevation Uintas remain in the "rain shadow" and bear the brunt of howling gale-force or stronger winds. All this creates a shallower, less stabile, more unpredictable snowpack, or what Gordon characterizes as a "continental snowpack."

"It's counterintuitive, but shallow snowpacks are fragile and spooky," Gordon says. "Once it starts snowing or blowing, we often see tree-snapping avalanches break to the weak, sugary layers near the ground." The Uintas, being higher than many other Utah ranges, experience colder temperatures around the peaks that preserve weak layers beneath fresh snow like that arriving on this night. The Uintas invite superstition: Gordon calls the peaks' reputation "a sketchy mystique," a place where for many years skiers were reluctant to go, fearing they would vanish into an avalanche and never be found.

Gordon's building anxiety tonight centers around the forecast he'll have to post online in four to five hours. The advisory that will likely be read by hundreds of people who might be considering venturing into the Uinta backcountry—skiers, snowboarders, and snowmobilers, the latter a growing constituency of Gordon's. Gordon and his fellow forecasters at the avalanche center are like sentries to the resort ski areas and Utah's backcountry, the out-of-bounds as it's called by skiers and ski patrollers. Each day of the winter recreational season, they post advisories for all of Utah's mountain ranges, from the northern reaches of the Logan and Uinta Mountains down to the state's prominent southern range, the La Sals, which are known for being avalanche prone.

In their advisories, forecasters like Gordon capture the unique characteristics of the slopes and threats posed by constantly shifting moods of snowfall, altitude, temperature, terrain angle, and wind. Not least of all, forecasters seek to chronicle the particular hidden history of individual snowpacks. In certain spots a lethal force can hibernate, awaiting the trigger of a skier's presence.

Gordon, who prides himself on discipline and a strict regimen, feels an even greater irritation as the clock ticks toward 11:00 p.m., depriving him of needed sleep as he will have to rise at his usual 3:00 a.m. to work on the day's forecast. His heightened anxiety has been at least two days or more in the making. The prior weekend, a large snowstorm preceded the one now haranguing the peaks; with few exceptions since November, a cycle of storms has created an "atmospheric river" of precipitation that, in the cold, winter skies of northern Utah, has translated to historic snowfall. The only place outstripping Utah are the Sierra of California, where specific storm snowfalls have been tracked not in inches but in feet.

The past five days illustrate what Gordon thinks of as "the true ferocity" of winter. As he paces his darkened home, Gordon calculates that the two storms will have shed sixty inches—five feet—of snow, which he visualizes as huge slabs of fresh snow on the faces of resort ski runs as well as in the ungroomed, unpatrolled backcountry. Compounding on vast slopes, the new, deepening snow is adding tons of pressure on the aging layers beneath.

"The whole time my head is churning," Gordon says. "What will tomorrow bring? We are deep into an intense storm cycle. Your gut tells you something is going to happen. There's 'close call, close call, *boom*—accident.'" He has received specific alerts, information culled from weather data sensors, and evidence sent from his wide network of avalanche spotters—skiers, patrollers, and backcountry guides who work in snowcat and heli-ski operations. That very day, he has been texted three photographs by an avalanche forecaster embedded with a snowcat skiing company in the Uintas, a spectacular collection of steep slopes known as Thousand Peaks Ranch.

As he studies the photos, Gordon sees the aftermath of a natural avalanche, distinguishable from one triggered by a human presence. The specific place, called Four Eagle Bowl, is one Gordon knows well. Angling about thirty-six degrees from the peak, the ridgeline and bowl fanning out beneath it had acquired at its zenith an immense natural cornice—a cantilevered ledge of snow sculpted by the wind. Hanging well out over a potential ski slope, the cornice, to Gordon's experienced eye, would be like smoke before flames suddenly explode from the high windows of a skyscraper. Once a cornice ruptures and falls to the face below, its weight is frequently the tripwire that sets off an avalanche with the power of a localized hurricane.

Reviewing the photos of avalanche debris, Gordon understands them to be "a great indicator of bigger things to come, a portent of other activity." Even more alarming is that he knows Four Eagle Bowl is not on his list of high-avalanche danger spots, so he receives the photos as more of a whisper than a cry: "This thing is tapping me on the shoulder."

Exhausted by worry and his long day, Gordon climbs into bed and tries to sleep. By 2:20 a.m. he wakes not to an alarm but to his own discomfort. Whether driven by the remnants of bad dreams or the seething snowstorm outside, he gets up, careful not to awake Anita, his wife of sixteen years. Although Anita has become accustomed to her husband's quiet, nocturnal work, she knows that his daytime forays into the backcountry are not without danger. "We make sure we kiss each other bye," she says, "and say 'I love you.'... He's great about texting me and letting me know where he is."

Making his way to the kitchen, Gordon pauses to check on Baby, his three-year-old gold-capped Conure parrot. Baby and Gordon share a relationship that brings the reassurance of simple affections, even a kind of peace, to a life where his job means constant obsession with violent storms, skier injury, and the ever-present possibility of having to dig up the dead.

Gordon jump-starts his morning as always, fixing "an amazingly powerful cup of espresso." He tops it off with Folgers instant coffee grounds and a dash of soy milk. A vegan, Gordon is as strict about his diet as he is about his daily weight-lifting workouts; that discipline extends to how he sifts through early morning reports to begin constructing an avalanche advisory. In the dim light, he switches on his computer. Sitting before the glowing screen, he reviews weather data from an array of sensors positioned throughout Utah's northern ranges. First, he looks at a series of remote NOAA devices planted at key spots in the Uintas. Gordon has placed six sensors himself. Of those six, three are located on "wind sites," ridgelines where anemometers spin out readings for wind speed and direction; each hour the instruments report averages of speed and maximum gusts. Today he sees winds from the south and southwest steadily blowing thirty to forty miles an hour, gusting into the seventies. The scene Gordon constructs in his head is one of dark, inhospitable peaks where storm winds snarl, whipping deep snow.

Temperatures are recorded by the wind sites, frequently well below freezing in winter. On this morning the prevailing temperature is 16 degrees Fahrenheit, with a wind chill of minus 6 degrees Fahrenheit. Gordon mentally pieces together the situation in the Uintas, which only stokes his concerns. He sees a mountain environment where Mother Nature is on the equivalent of a drunken tear. Then he examines reports from the other three sensors: snowfall and depth instruments. In addition to the forty or so inches of snow from the past weekend, another fifteen inches has fallen during the night. And the forecast is for snow to further dim the faint light of day.

One other array of sensors draws Gordon's attention. SNOTEL reports record the depth of snowpack as well as the water content in snow produced by various storms. Originally, SNOTEL data was intended for use by watershed monitors, people who track and predict snowmelt that filters down mountain faces into vast drainages that fill alpine streams and lakes. This data enables the monitors to estimate water levels and expectations for reservoir depths in the spring, summer, and fall—critical information for both farmers and dam managers planning to avoid floods.

SNOTEL sensors have become another key tool for avalanche forecasters, and as Gordon reviews them his alarm spikes. Apart from the amazing amounts of water being deposited onto mountain shoulders by the snowfall—the weight of which stresses the snowpack even more—Gordon sees that all the recent storm forecasts have fallen short and that the storms have "totally overproduced." This means so much snow has fallen, it is distributed in heavy amounts virtually everywhere. Very localized storms will often favor a certain area, and create the extremes, the spots that stand out to even an unpracticed eye as places where avalanches are likely to lurk. The huge cornice that crashed down into Four Eagle Bowl instigating an avalanche is such an extreme.

"As humans, we don't always pick up on the nuances," he says. "We focus on the extremes. What I'm interested in are the winds loading up the ski starting zones. I'm interested in *all* the fresh, white paint." He calls it "paint" because it conceals flaws in the terrain. Under such heavy cover, he says, "things start to get connected, you lose terrain boundaries and the slope becomes planar. . . . That's when things start to get super spooky."

WHEN GORDON CONSIDERS THE FULL picture in the early morning of Wednesday, January 11, he summarizes all the Uinta range data. "I've got seven inches of water and like sixty inches of snow," he says. "For that mountain range those are unheard-of numbers. As I'm looking at this I'm thinking, 'The shit is hitting the fan.'" Avalanche forecasters, as a breed, like certain warriors, police detectives, or captains of passenger ships, are a careful, discerning bunch. Crying wolf is a rarity; being taken seriously by backcountry adventurers is both an earned, professional mantle and a point of personal pride.

As 4 a.m. approaches, Gordon finds himself under the pressures of the clock and also that of his mind, which is working against his customary restraint. He puts together an avalanche forecast with all the typical observations culled from the data he's reviewed. But he knows the most critical piece of the advisory—the headline skiers and boarders will look at before reading the detail—will be the danger level of the category.

The international scale for avalanche threat classification is not unlike the Saffir-Simpson scale for hurricanes in that it strives to gauge probability and severity. Avalanche threat can fall into one of five categories—low, moderate, considerable, high, and extreme. Each is color-coded: green at the genial end of the spectrum and black at the catastrophic end. "It's a huge mountain range and I have to consider many variables and terrain nuances," Gordon says. "While I err on the side of conservative, I try to explain in laymen's terms what's going on and what the consequences look like if you do trigger a slide. Bottom line? . . . I tell the truth and try not to overforecast. But I knew this was the time to get peoples' attention and not wind up on the five o'clock news."

He is strongly inclined to put a "high" rating on the avalanche threat and back it up with pointed conclusions from the alarming data. He sits there, staring at the computer and all the information he's entered into the Utah Avalanche Center's website for the Uintas daily report. "I'm just about to hit publish and go to the gym. And I stepped back." Getting up from the computer, he paces the kitchen and has a last sip of coffee. Sitting down again, Gordon changes the description of the advisory: the danger "is HIGH and might be bordering on EXTREME." At 4:06 a.m. he hits send. A moment later, the warning hits the avalanche center's website with the authority of a police cruiser's flashing red lights.

"Something felt different to me that day," he says. "I knew the danger was high. In seventeen years, I've only issued two extreme ratings, this being one of those days." Less than thirty minutes later, he leaves for his weight-lifting session at the Cottonwood Heights Recreational Center. "I left feeling really solid about my forecast," he says.

So solid, Gordon thinks to himself, he will not even need to travel to the Uintas that day to visually verify the details of his advisory. No one, he thinks—and hopes—would be likely to go out there into the white maelstrom. After all, he has prefaced his posting: "Heads up. . . . It's game on and this is the real deal. Once triggered, today's avalanches will break deep and wide, resulting in a large, dangerous, and most likely unsurvivable avalanche."

3

THE DRAGON AWAKES: "DEEP AND WIDE"

In the dying light of day, with snow still punishing the Uinta peaks, Utah snowboarder Jeremy Jones is falling. From the higher margins of the slope, his companions see him and they cannot believe what is happening. Even if they thought to scream, Jones could not hear them. Wind, and the sound of tons of snow and ice ripping through everything in its path, deafens him. As the convex roll of the mountain comes up at him, Jones disappears from the snowboarders' view. On high, a strange silence settles over the ridgeline. It's like the remaining riders are stranded, helpless, lost at sea while one of their own has vanished in a huge wave.

Below them, Jones fights for his life. He's struggling to stay on his feet, to retain some steerage of his snowboard. He fights to force his body to the right side of the slope, a blur of trees where he might escape the avalanche dragging him down the featureless face of the mountain. The slide is like a net he's ensnared in. It swarms all over his body. And then his feet run out from under him. The snowboard is like a plow blade caught downslope in a landslide. It anchors his legs deep within the violent whirlpool.

For a second, Jones feels himself angle a fraction to the right, and he thinks he might have a shot at "reachable freedom." Then a little wave of snow bucks him up from behind and flings him forward. His bid for the trees fades with the wake of the slide. He resorts to frantically slashing his arms at the turbulence around him, swimming to stay as much above the snow as he can, fighting for breath. Miraculously, he sees a small tree ahead. The slide hurtles him right at it. Jones raises his arms instinctively, hands splayed as if they are the only thing between him and certain death.

He catches a branch jutting downhill, his palms grasping just past the trunk. He feels his hands being ripped the length of the branch, breaking away needles and young shoots of pine. Then he feels only air between his fingers. The avalanche has sucked him back in, as if the tree was only an illusion of rescue.

"Death's not popping to mind yet," Jones recalls remembering in that moment: "You're not going to die on your kids today." His grasp of the branch has slowed his fall a little. He's plummeting from the rear third of the avalanche, still sliding but clinging to a sudden hope that the nose of the slide will crash into an obstacle and the rear will accordion into it, slow itself, and then settle with his body floating on top.

Just as Jones finishes the thought, he feels a sick tingle in his stomach. The second convex roller he slid over in his earlier run has now come up. The rear of the avalanche is like a wave peaking over the roller's sandbar. He feels the sudden punch of the roller's steeper pitch. Then he's rag-dolled up into the torrent of snow that breaks in a wave above his head. The slide slams his body against the rock-hard trunk of a tree. "As quick as I hit the tree and sat down, the slide stopped," he says. In the confusion of the moment and shock of impact with the tree, Jones passes out for a second or two. But then he's suddenly alert. He doesn't even sense pain from where his snowboard and legs have been smashed by the tree. "Adrenaline was so high at that moment."

Jones senses people bounding down the slope, coming to rescue him. Some seem to be below him. And then he hears them shout. "We're missing Mike!"

Somehow, Mike Nelson had dropped into a run off to Jones's side. Riding a snowboard without bindings, a kind of mountain surfboard sometimes called a "no-board" that is leashed to the rider, Nelson disappeared in the slide. Yells ensue for people to switch their avalanche beacons to "search" mode. Jones, dazed by his wild caroming down the mountain and the tree's battering of his lower body, confusedly starts to feel for his beacon. Although he can barely move, he knows that if his beacon remains on "send" it will confuse the searchers looking for Nelson. Finding the beacon still lashed to his chest, he switches it to "search."

Then rescuers are at his side. He has managed to unlock his right foot from his snowboard binding, but it's "not there, it's just flopping." He shouts: "Find Mike!"

Already, five minutes or more have passed since the avalanche crested on the roller and stopped. While no one pauses to say it, the snowboarders have had enough exposure to avalanche training to know about the ten-minute window for finding a buried rider. Much past that and they will be searching for a dead body.

DOWN THE SLOPE, SETH HUOT scans the debris field for any hint of color. In the dropping light and snowy sky, it's hard to discern anything other than a massive confusion of snow. It's like the avalanche's anger has dissipated here, and the field before Huot is hushed and gray as twilight hovering above a graveyard.

Huot covers ground fast. His beacon cradled in his hand, he tries to lock in a signal. Unbeknownst to him, the device is of no use. Earlier in the day, Nelson saw his beacon's "low battery" light click on and he switched it off. As Huot boot-packs his way through the deep snow and glances up from the beacon, he spots something. Off to the right, in the thick of the slide's main rubble, is a thin edge of Nelson's board, barely a fingertip of fiberglass poking through the snow.

Rushing over, Huot discovers a small stretch of the leash that was tied to Nelson's leg. He grabs it and begins to pull. There's weight at the other end, and Huot tugs harder until he reveals a portion of Nelson's leg. He realizes Nelson's head is pointing downhill. Swiftly, he clears snow where he judges Nelson's mouth is covered. Six inches under, he finds the man, his face shaded blue. Nelson has been submerged for about twelve minutes. Huot quickly frees him and Nelson coughs his way to finding measured breath again.

If the boarders are given a moment of comfort that everyone is alive, the enormity and urgency of the situation before them starts to weigh as sure as the cold and coming dark creep around them. Whatever margin for error they might have possessed, they know it has been used up in the avalanche accident itself. A realization dawns that might have seemed trivial that morning: they are well beyond any cell phone service.

Jones's circle of riders inventory the tools and tasks at hand. In the snowcat they have a machine that's the equivalent of a life raft, but they know it is slow, cold, and carries a checkered operational history. The only other option is to attempt to hike out. But the prospect of hiking for hours in deep snow and diving night temperatures feels like a suicide mission. Moreover, they know Jones is physically and mentally strong, but they don't know how extensive his injuries are. He could lapse into shock, or worse.

Brock Harris, one of Jones's core group, goes to work on Jones's legs. His right one is certainly snapped above the ankle, his left at least seriously fractured. When he hit the tree, the left side of his body bore the initial impact, and he may have cracked ribs or sustained other internal injuries. Using avalanche snow probe poles, Harris constructs splints for Jones's right leg. He lifts Jones from the snow and the two men try to walk toward the trail where the snowcat sits. After fifteen feet, Jones can't bear any more weight on his left leg.

Taking some cordage, Harris aligns Jones's snowboard boots, then wraps rope tightly around them so he isn't forcing weight onto a single leg. Harris slides Jones onto the deck of Nelson's no-board, and with the leash he slowly lowers Jones down the angle of the snow slope, six feet or so at a time. Tedious as the rescue is, Harris finally conveys Jones to the snowcat, where he is helped inside.

"I knew we had a long road ahead of us," Jones says. "I was wet, I was cold. My core was warm though. And I trusted my friends." Since the morning's initial gathering of the group, Jones has been reassured by the presence of his close friends. Now, he and Nelson come together in the cat, not just as old friends but as fellow survivors. The moment is awkward for the other riders, a little surreal for both men.

"I'm stoked you're alive, man," Jones tells Nelson. Nelson, stunned from having been buried alive and by the reality of their situation, manages similar sentiments. He lapses into a contemplative state as Jones, despite his injuries, tries to remain engaged with the group. As if he's still the leader, as if by displaying confidence he can reassure the younger riders that things aren't as bad as they may seem.

But it's hard to reconcile Jones's battered image with any sense of relief. There's additional preparation for the long transit to the trailhead. To make sure the cat's lumbering motion on the trail doesn't shake Jones's broken legs excessively, duct tape is tightly wrapped around his boots, immobilizing them as a makeshift cast. Pulling some extra base-layer clothing from his pack, Jones pulls on a dry shirt and one of the drivers loans him a coat. The engine noise jackhammers the cold cabin, the machine lurches forward, and in the dark they begin a slow trek back.

In his head, Jones begins another risk talk. "There's high risk and there's unknown risk," he says. "We got to the cat and the unknown risk starts to fade. But we're always at high risk." The day's unseen risk—the surprise breaking of the avalanche on a face they had already boarded across—has passed. But Jones knows there are a hundred things that can go wrong in the three hours he estimates it will take the cat to make the trailhead.

Looking around the cabin interior, he sees expressionless faces, people who have retreated into their own worlds of reliving the accident, guiltily wondering if they did anything wrong, if they missed some sign of the incipient slide. A trip that started out that morning on a note of excitement, and a feeling of privilege, has instead turned into a morbid journey back to a world that will have more questions than most riders want to answer. With the monotonous drone of the cat engine in their ears, the riders slip deeper into themselves.

AFTER NEARLY TWO HOURS OF progress, they meet their first obstacle. Just a few hundred feet from the clearing where they had stopped for morning beacon practice, the engine dies. The breakdown leads Jones to do some quick mental math. He figures it will add at least two to three more hours of being stranded in the backcountry. Whether the drivers get the cat going again, or a small party is forced to hike out and alert search and rescue, either way he is looking at a longer, colder night, the pain from his broken bones growing more pointed by the hour.

After some discussion, it's decided that three riders will hike the remaining two miles to the trailhead, drive a car into Oakley, and phone Summit County Search and Rescue. The snowboarders find at least two feet of fresh snow has fallen on the trail that day, on top of the already more than four feet from the weekend. With headlamps glaring, they punch through the drifts, disappearing in the dark like bare survivors who may never be seen again.

It's about 8:00 p.m., and Jones lets his thoughts run to his family and the friends who saved his life. He thinks back to all the years he and his core rider group practiced in the Brighton backcountry, their creation of the small survival kits that Harris drew from earlier when binding Jones's legs. "It was a phenomenal display of knowledge," he says, "once the shit hit the fan." For the first time of what will be weeks of a daily mental replaying of the accident, Jones sifts through the day, through the precautions they took on the mountain, and whether there were signals from the surroundings, increments of wariness he might have overlooked or disregarded. When he plays the film in his mind, like those films he and his friends made back in the day, he sees himself on the first run down the mountain with no evidence of avalanche threat. Maybe, he thinks, after that run complacency took hold, that and the desire to get in one last run for himself and the group.

"I sort of let that slip more than I should have," he says. But even that sharpening of his focus on a single judgment point doesn't persuade Jones he would necessarily spot the cocked trigger of unknown risk next time. "In hindsight and regrets, there are none," he says.

As the hours pass, the three riders have made their way into Oakley and, with cell service restored, have contacted Summit County Search and Rescue. As the rescue team plans a mission into the backcountry on snowmobiles, an alert goes out to other Salt Lake–based responders. In the air above the city's Primary Children's Hospital, a LifeFlight medical emergency helicopter has just taken off and they pick up on the rescue notification. Not knowing the full extent of Jones's injuries, the chopper crew concludes the situation sounds dire at best, particularly given that the avalanche victim is marooned at night in the freezing backcountry. They know it will take time for Summit County rescuers to mobilize and travel on snowmobiles to the site.

The LifeFlight crew decides to fly in over the Uintas and see what the situation looks like from the air. An Agusta Grand 109, the chopper is designed for high-altitude flight, originally built for rescues in the Swiss Alps. It bears a collision avoidance system for navigating above mountains, a bundle of "Highway in the Sky" piloting instrumentation that guides pilots in low or zero visibility.

Even inside the snowcat, Jones and his companions hear the distinct, twin turbine sound of the heli coming in over the Uinta Range. When it grows closer, they realize it must be part of the rescue operation; the heli eventually hovers above the clearing and disabled cat. From its underbelly, a powerful searchlight pierces the falling snow and illuminates the ground below like the accident scene it is. Then the heli drifts down toward the clearing. Within minutes, the heli crew is inside the snowcat with a backboard. Two men and a woman question Jones about his condition and he offers a laconic "I'm good." The woman inserts an IV into his arm; she tells Jones she will give him an injection to kill pain and knock him out during the flight to the hospital.

Once airborne, the LifeFlight heli crew find the weather deteriorating. The snowstorm's full rage has come back in as if to wield its final vengeance. All ground reference has been lost and the crew is put into a twenty-minute holding pattern. Soon, one of the crew spots the bare, distant headlights of a vehicle weaving its way through nearby Provo Canyon. They track the vehicle's progress, knowing it will eventually lead them near Timpanogos Regional Hospital in Orem.

Inside his head, drowsy from medicine, Jones continues his risk talks. He's having worries about the helicopter navigating in the storm, about the holding pattern, and the search for a hospital with a clear landing site. He can't help himself. It's who he is.

When Jones would come home from all his days of backcountry training and survival rehearsals near Brighton, he would tell his wife about them. "'You're crazy,'" she'd say. "One day it will pay off," he'd respond. Injured as he is now, Jones believes it did pay off. "If you can outlive one, you're doing something right."

If you're not prepared to fucking survive—you won't.

IN SURGERY THE NEXT EVENING, Jones's broken legs are reset. He leaves the hospital in a wheelchair and begins the road back to the physical shape he'll need to go into the backcountry. There's no doubt in his mind he will get there. But maybe he's slightly more cautious about the world than before the avalanche. Nearly two months after his surgery, he's on crutches and decides he and his family need a break from the avalanche's aftermath. It's like they all have been living within the echoes of its thunder crack, triggering a massive wave of savage snow. So they drive down to southern Utah, to Saint George where there's sun and congenial bike paths for his wife, Sher, and their kids, Adi and Cru.

At the start of a recommended bike path, Jones finds his mind drifting back to the Uintas trailhead, to the beginning of that hallucinatory day in the backcountry. He feels a faint, inchoate fear rise inside him. Injured as he is, he can't ride the four-mile trail with his family. And he's forced to admit that if something happened—an accident—the best he could do is helplessly hobble on his crutches two miles in.

"I had this thing happen that makes me look at things differently," Jones says. Still, he doesn't shy from the prospect of returning to the out-of-bounds and even to the mountain slope where the avalanche swept him away. When Mike Nelson had been rescued and helped to the snowcat to regain his wits, Nelson realized that a GoPro camera he'd had strapped to his body had been ripped away in the slide and lost. Jones plans to go back out there when the snow melts, in the late spring or early summer, to see if he can find the camera and the video footage it may hold. It would be something to see the point-of-view footage of the avalanche that caught him and Nelson.

Jones thinks about the human factor of the accident. Though he bore the worst of the injuries, he will mend. Nelson skirted death but came back, like Jones, with insights about the out-of-bounds—and life. Seven other riders survived. “This for nine peoples’ lives?” he questions, balancing on crutches and looking down at his legs.

“A pretty fair trade. A pretty fair trade.”

4

CRIME SCENE INVESTIGATORS OF THE OUT-OF-BOUNDS

After filing his avalanche advisory early that morning, and putting in his usual two hours at the gym, Craig Gordon spends much of his Wednesday tapping into an extensive network of sources for reports from the backcountry. A big part of his forecasting job is to collect, on any given day, episodic, real-time missives from skiers, snowboarders, and snowmobilers who are traveling in the out-of-bounds or who have just returned from a day there. Emails, texts, and digital photos come his way each day from a wide array of contacts he's made over the years. While most of the data arrives digitally, as does information from all his electronic weather sensors in the mountains, Gordon knows he would have a blind spot if he didn't also get firsthand reports and newsflashes from skiers. Out there, they come upon fractures in the snow, witness slabs careening down like huge snow toboggans, and see other signs of the avalanche dragon waking.

Despite the extreme warning he issued that morning, Gordon's day is almost eerily uneventful. By early evening he's back home in Cottonwood Heights. Around 7:00 p.m. he gets a phone call from a member of his reporting network. The caller has a question specific to the Uinta range.

"Hey, do you know if the powder cats are down and clear?" the caller asks. "A friend of mine's wife called me and said her husband and some guys went cat skiing and should have been home by five." He mentions the name "Jeremy Jones" and at first Gordon mistakenly thinks it's the other professional big-mountain snowboarder who shares the same name. This other Jeremy Jones has an eponymous snowboard company in Truckee, California, and makes ski and snowboard films with his brothers, the owners of Teton Gravity Research. The caller dispels Gordon's confusion.

Gordon thinks the caller may mean Park City Powder Cats, a guided backcountry concern of four snowcats that operate in the Uintas. Gordon knows they roam a vast but private range called Thousand Peaks Ranch, and they typically finish their skiing day by 4:00 or 4:30. He remembers a conversation he had some months earlier with people operating a single cat in the Uintas.

Gordon's mind races, as if he's seeing an aerial view of specific Uinta peaks. Before becoming a forecaster with the Utah Avalanche Center, he worked as a heli-ski guide and pioneered some helicopter-accessed terrain in the Uintas. Months of scouting peaks where a helicopter could land have embossed in his mind's eye a mental, topographic map of the Uintas.

First Gordon thinks of an area called Chalk Creek drainage and a zone he calls Wallyworld, a lower-elevation canyon surrounded by avalanche funnels which he refers to as "massive avi paths." He imagines a scenario there where a snowcat has strayed into a terrain trap, a gully under high peaks, where a deluge of descending snow can bury a snowcat and all its passengers. His mind goes to another prominent destination for summer hikers and winter recreationists—the Smith-Morehouse Reservoir. Summoning an image of it, he fixes on a "big amphitheater of mountain peaks that drain into the reservoir."

Given his advisory that day, either scenario leaves Gordon highly alarmed. Worse, the people who told him about the single cat operation also allowed that the machine has had breakdowns; the owners had once been forced to abandon it in the middle of the night and return later on snowmobiles to repair it. When Gordon hangs up the phone, he's back on the time clock. "I don't exactly know where these people are," he says, "but I have become part of the telephone and text string. And I am accountable." Aside from his professional obligations, he's "a little pissed." Of the still raging snowstorm, he says, "I know what's going on out there. You're always waiting for the shoe to drop. . . . An accident, a death. I made the effort to let everyone know how sketchy it is."

Sorting through his list of contacts, Gordon decides to phone Tyler St. Jeor, a Canyons ski resort patroller who also works with Wasatch County Search and Rescue. Gordon has seen more deadly avalanche accidents than he wants to recall, but he's also seen plenty of search parties called up and sent into the mountains at night when someone was perfectly safe and just a little late getting home. He's looking for St. Jeor to hear him out and then weigh in on possible next steps. "I didn't necessarily want to get the cavalry out yet."

St. Jeor knows the reputation of the Uintas, and he is as plugged into the weather of the last four or five days as Gordon. "If I was them I'd get a helicopter in the air," Gordon hears St. Jeor say.

Now it's past 8:00 p.m. and Gordon sorts through what little he knows and what he doesn't know. No one has mentioned a phone call from the snowboarders, which tells him they are likely still deep in the backcountry without cell service. That alone, knowing of the breakdown-prone snowcat, tells him "something worse has happened."

His phone rings again. It's Trent Meisenheimer, a colleague of Gordon's at the avalanche center. As word has gone out over the far-ranging network of patrollers, forecasters, and rescue people, Meisenheimer has talked with Sher Jones about her husband. She's just received word of the accident, the helicopter rescue, and Jones's injuries. Meisenheimer and Gordon decide they'll take snowmobiles out the following morning to investigate the avalanche. Just as detectives would scour a murder scene on a city street, they are bound to assess the aftermath of the slide and try to understand its origin. Knowing what caused it won't help Jones's legs heal faster, but there may be clues that will help in their forecasting in coming days and save other backcountry boarders from a far worse fate.

WAKING EARLY TO POST HIS Thursday, January 12, advisory, Gordon finds his cell phone's "texting telegraph" lit up with details of the avalanche, the wayward snowcat, and Jones and Nelson's rescue. Gordon will send out a report nearly identical to the previous day's, referencing the accident: "What we do know is two snowboarders triggered a large slide, were buried, recovered, and sustained very serious injuries, but at the end of the day survived."

Mentally he gears up for the day ahead. He'll have to be available for some early morning media before heading to the Uintas. As part of his job, Gordon is the main media contact for the avalanche center and does regular on-air reports in the winter months. Backcountry accidents mean he's on call to the Salt Lake TV stations as well as the area's newspapers. So many Utah residents ski, snowboard, and snowmobile that the report of an avalanche accident is similar to news that a kidnapper is roaming northern Utah. An avalanche accident in the Uintas means there could just as easily be one brewing in Little Cottonwood Canyon where ten-foot-deep avalanches often close the roads, or in the Ogden Valley, or just above Main Street Park City. It's cause for heightened caution and, as local news goes, it's invariably the story of the day, something discussed in grocery store aisles, office hallways, hair salons, and brewpubs.

"Investigating an avalanche accident is like CSI," Gordon says, referring to the TV show about crime scene investigators. If the gravity of the situation may seem slightly less, the unfolding narrative of the accident demands that Gordon and his fellow forecasters use many tools that are remarkably similar to those of CSI detectives.

By the time Gordon and Meisenheimer reach the Smith-Morehouse trailhead, they are equipped like men going into battle. They have trailered a snowmobile behind their vehicle and quickly unload it to speed their journey out to the avalanche debris field. It's a bit under five miles, and two miles in they come upon the stricken snowcat. After Jones's evacuation, the snowboarders and cat drivers were rescued in the night by a Summit County Search and Rescue party on snowmobiles.

Knowing that some of the snowboarders almost certainly read his warning the previous day, Gordon admits to mixed emotions, feelings "of compassion and being pissed off." Arriving at the sight of the slide, he is bowled over by the sheer volume of the stagnant jumble of snow and ice. He and Meisenheimer record the avalanche's basic dimensions. It's three hundred feet wide across the slope, a football field. Above—where Jeremy Jones rode the slide in its rear third, down to where it buried Mike Nelson, and past that to where debris spilled onto the snowcat trail—it's eight hundred feet of devastation. The depth of the avalanche is four feet.

For Gordon and Meisenheimer it's not hard to imagine the weight and momentum of the slide—three football fields end to end, loaded with white cement up past an average person's waist, unleashed down a nearly forty-degree mountain face, inclined halfway to almost pure vertical. It's akin to the havoc resulting from a runaway locomotive dragging fifty to a hundred railcars behind it.

Gordon thinks of his advisory from the previous morning. "Deep and wide," he'd described the likely avalanches. His prediction sits before him, a manifestation in jagged snow. Using skis with climbing skins, he and Meisenheimer navigate around the snow to perform more tests. They've brought a video camera and a drone, which they launch to capture aerial footage of the slide zone and impart the huge, geographic scope of the accident. They carefully dig a snow pit going three feet across and nearly five feet deep, just below the suspect snow's weak layer. Like CSI detectives, they find the hidden clue to the slide's break.

At just about the middle of the snowpack—four feet, the thickness of the avalanche—is a thin layer of what Gordon calls "faceted snow." It's sugary, rotting snow in which the snow crystals are poorly bonded to each other. Loaded with all the weight from the past four days' storms, the snowpack at that spot has cracked, releasing the layers above to answer the pull of gravity and human weight suddenly borne by it.

Gordon assembles a chart of the amount of water that's been added to the snowpack in the past four days—seven inches. He deems this measurement "a truly remarkable amount of water weight." It's difficult to precisely gauge the weight of an avalanche due to the mix of snow, ice, rock, and tree fragments. But at eight hundred feet long by three hundred feet wide and four feet deep, the slide propelled about 960,000 cubic feet of frozen debris down the mountain. In Utah, avalanche forecasters calculate the weight of snow slabs at about 70 percent air and 30 percent ice, or a volume of 288,000 cubic feet in this instance--slightly more than 8,900 tons. The weight Gordon is confronting is roughly equivalent to that of thirty-seven Boeing 747s. Enough down force, in other words, that the avalanche would flatten anything in its flight path.

"No matter how strong the existing snowpack was," Gordon concludes, "even the slightest weakness will reveal its cards under this kind of load." He thinks of Jones and Nelson, caught in the horrendous turbulence, rocketing down the mountain face. "This thing is fucking smoking, packing heat, bursting trees. They are like toothpicks in a hurricane. You are just in awe of the power of nature."

Standing, Gordon cocks his head upslope as if he can hear a replay of the avalanche. There's the thunder crack of the break, and then its roar, nearly deafening as it gorges itself on more and more debris, splintering trees, raking loose rocks into white jaws. He imagines when the slide suddenly comes to a stop. The silence has its own wavelength of sound. It's the barely brushed drummer's cymbal, a building suspense, against which all other sounds are punctuated. The frenetic cries of the unhurt snowboarders, sprinting down, yelling about beacons and probe poles.

Then there's Jones and Nelson. He imagines Jones being first stunned and then elated that he's alive. He pictures Nelson under the snow—a sight Gordon has conjured hundreds of times in avalanche awareness courses as he describes a burial and the slow onset of death. He sees the man's face, an ice mask forming before his mouth and nose as Nelson's labored breaths discharge more and more carbon dioxide into the small sphere above his head. The man tries to move, but it's like his arms and legs are lashed down. A weight has settled onto his chest and his lungs fight to find room to fully expand within his ribcage. The cold of the ice mask begins to turn the man's cheeks and forehead the blue of constricting veins.

Gordon forces himself back into the world at his feet. With more observations recorded and photos snapped, the two investigators prepare to load their gear on the snowmobile and head back to the trailhead, knowing they'll pass the broken-down snowcat, a sad reminder of all that went wrong in the Uinta backcountry. Scooting through the trees on the same path the cat had traveled, Gordon can't help but start assembling all the data in his head, urging himself toward a conclusion about the accident.

"Like any avalanche accident, it was a series of events," he says. "Like an aeronautical accident, it's a series of things and events that lead you so deep down that road you can't turn around." His advisory had said anyone caught the previous day in an Uinta avalanche would find it "likely unsurvivable."

But there's one consoling detail: today Craig Gordon knows that Jeremy Jones and Mike Nelson are *alive*.

5

THE SNOW DETECTIVE INSTRUCTS

It's a Thursday night in January and a conference room in Ogden, Utah, is packed with about thirty-five people, mostly men in their twenties and thirties who have assembled for a two-day workshop called "Introduction to Avalanche." The conference room has been loaned to the Utah Avalanche Center by the Ogden-based Amer Corporation, makers of Atomic and Salomon skis and the parent company for the high-end adventure clothing maker Arc'teryx.

The mood and look of the crowd is distinctly that of outdoor bro culture. Slouched in their chairs, the young men—and two young women—wear loose athletic pants and down vests. Many foreheads are nearly obscured by beanies pulled to eyebrow level at the edges of which telltales of hair dangle, indicating the skiers and boarders are long overdue for haircuts.

The two-day avalanche class is structured for this first three-hour session in the classroom to be followed two days later by a full day in the so-called side-country of Snowbasin ski resort; side-country is adjacent terrain not typically patrolled by the resort, but not considered completely remote. It's often the place where first steps are taken to learn travel techniques for the backcountry.

Snowbasin is the site of much of the 2002 Winter Olympics ski racing and it's a choice location for an outdoor classroom in which to learn avalanche assessment skills. The upper mountain is adorned with steep couloirs or "chutes" where avalanche threat is aggravated by every new snowfall. Many mornings, the sonic booms of avalanche mitigation explosives echo across the Ogden Valley below the resort.

About thirty minutes before the start of the class an avalanche center staffer makes the rounds, surveying who has avalanche gear—beacon, shovel, probe pole. Some proudly say they already are equipped. Whatever gear participants lack, the avalanche center will try to round up for the long Saturday in the snow.

Now it's just past 6:00 p.m., the assigned start of the class. As the clock ticks away, the room falls silent and the mood takes on an anxious feel as if the group of adrenaline junkies are being denied any way to release their pent-up energy. At 6:15, Craig Gordon strides into the room. A tall, lanky man still dressed in his work outfit of black ski bibs, Gordon also bears an avalanche beacon strapped to his chest. Under the shoulder straps of his bib, he wears a maroon-colored fleece with a Utah Avalanche Center logo on it and the stitched words "Keeping You On Top."

He apologizes for his lateness. "I've just come from the Uintas," he says, "where we were investigating the accident that happened yesterday up by the Smith-Morehouse Reservoir." He has a striking presence. He's lean as a leopard and equally muscular, his arms and legs bearing muscles that are more extruded than bunched. Above his tanned face, his black hair is pulled tightly back and crowned by a Bodhisattva's topknot, tightly wound, the hair braids like a trim carving in obsidian atop his skull. The expression in his eyes ranges between the seriousness of his profession and a kind of mischief that the outdoor bros can clearly relate to.

When Gordon talks about the science of avalanches, he could be a professor teaching a class at a small liberal arts college, articulate yet relaxed, revealing he's completely in command of his subject. But he peppers his remarks with words like "dude" and "stoked" that expose his New Jersey, surfer boyhood, which clinches the attention of the outdoor bros before him. It lets them know that at heart he's one of them.

One other subtext emerges as Gordon talks about avalanches: ultimately he's talking about racing to avoid death in the mountains. Spelling out the conditions that almost certainly generate natural or human-activated avalanches, he concludes with phrases like "and then you get smoked" or "you end up having a real bad day." The metaphor Gordon comes back to most is that of avalanches as "the dragon." Backcountry riders, he says, might "wake the dragon." Describing a specific storm, he'll say, "The avalanche dragon had a different characteristic." For Gordon, it's a potent image—the shuddering of something waking up, mean, volatile, hell-bent on killing the prey that disturbed it. *Dragons in the snow.*

On this night, surprisingly, he doesn't dwell on details of the Uintas accident. He's still processing its wide sweep of data and details; he needs time to study the still images and video. He hopes to interview at least one of the survivors to try to understand the "human factor" that played into triggering the avalanche; it might be a group dynamic that played down the day's risk, or an action by a single snowboarder. For now, Gordon sticks to the night's prepared presentation, which focuses on helping the group spot "obvious clues" of dormant avalanches and on terrain route finding that steers backcountry skiers around potential slides rather than directly into them.

One of Gordon's most notable achievements since joining the avalanche center is a program called "Know Before You Go," an interactive session for school kids that often begins with an attention-getting film. As the film rolls, spliced-in sections feature avalanches as the only on-screen performer, a kind of snow dramaturgy. These scenes leave audiences feeling jaw-dropping awe and a concomitant respect for nature's power that emphasizes the puny presence humans occupy in the out-of-bounds.

Classes such as the one Gordon leads tonight are a regular part of his proselytizing about snow safety and learning to use the tools of avalanche avoidance. It's been a long journey from his childhood on the New Jersey coast to the rarefied mastery of ski mountaineering he has achieved. Gordon's first exposure to skiing was when he was about four, and his father took him on a trip to Hunter Mountain in northern New Jersey. It was not love at first slide. He recalls the trip as a freezing "sufferfest" that involved a child's mechanical nightmare known as a "rope tow." Gordon's father had begun working part-time in the ski shop of the Swiss owner who had transplanted himself to New Jersey to preach the nobility of European ski culture. The evident romance of that, and a particularly helpful girlfriend of his father's, who coached Gordon on a subsequent ski trip, changed his opinion of skiing. "In a moment, I fell in love with it," he recalls.

Gordon's wife, Anita, says that as a kid Craig constantly drew pictures of ski lifts with him on one of the chairs. "It's what he was born to do." Skiing solely in New Jersey wasn't likely to turn him into an Olympic phenom, but a kid could dream, and Gordon remembers, in 1976, seeing his first issue of *Powder* magazine—a kind of skier's porn without the nudity. The issue contained an article about Utah and photos of sumptuous mountain slopes that served as his earliest exposure to "Know Before You Go" images.

"I thought, 'This place is badass,'" Gordon recalls. Three years after becoming smitten by those photos, he finagled his way onto a seven-day ski trip to

Utah. Paying $365 each, he and a friend landed at the Howard Johnson motel in Salt Lake City and began waxing their skis. Outside, the world of snow greeted them like pilgrims who had arrived at Mecca.

"It snowed and snowed and snowed," Gordon says. For the next three days, it snowed two to three inches an hour. At Snowbird resort in Little Cottonwood Canyon, he remembers dropping into the first morning turn of a run and being nearly submerged by snow, "just over the head." The next day, Gordon had what, in his personal origin story, is a moment that shined the way toward the rest of his life. He and his friend had made the first morning chair up at Alta ski area. They slid off the lift to the sudden sound of avalanche mitigation work above them, booms that echoed through the granite faces of Devil's Castle and other Alta peaks.

"It's a sunny, clear day, it's spectacular," Gordon says. "I hear the avalanche bombs going off." Standing there in awe of the freshly snow-doused peaks, Gordon and his buddy suddenly saw three or four Alta ski patrollers coming down toward them from the snow-mitigation work. In their black ski pants and distinctive blue patrol jackets bearing a large snowflake icon, shouldering heavy mountaineering packs, they looked like alpine gods. To Gordon they were beings instilled with not just the duties of protecting skiers from avalanches, but with protecting humankind from nature's wrath. And they all could ski like a dream—and get paid for it. Gordon turned to his friend and told him someday soon his plan was to return to Utah and join the ski patrol. "I thought, 'What a bitchin' job that would be.' The seed was planted."

AS HE APPROACHED HIGH SCHOOL graduation in the spring of 1980, Gordon had not wavered a jump turn from his vision. He had prepared to apply to the University of Utah, but his plans hit an unexpected mogul, a "speed bump in my life" he never saw coming. Gordon was diagnosed with a brain aneurysm.

"It was a life-altering event," he says. There was an operation and a long recovery; Gordon saw his plans begin to melt away like late spring snow. Partly it was his medical condition, but it was also complicated by his parents' aversion to his desire to head west, to a city where he knew no one, with the stated goal of chasing a ski patroller's jacket.

In 1984 he won admission to the University of Utah, packed up, and, like all westward-bound pioneers with grand delusions, landed in Salt Lake City. Initially buoyed by his teenage vision, Gordon saw his hopes immediately dashed on the reality of his shaky financial straits. He needed a job quickly, and the first one he secured was as a dishwasher. Like the dazzling image he retained of the ski patrollers, there was a hopeful aspect to his situation; his job at the Forklift Restaurant at Snowbird came with a prince's perk—a season ski pass.

If Gordon could rejoice in his ski privilege fortunes, his professors at school could offer no such benevolent feedback. Repeatedly absent from classes to ski, he found himself, by the end of the first semester, essentially failing out. "The guise of this college façade came through when my grades appeared." Managing to barely hang on into the second semester, he did not perform a turnaround with his professors. His delinquency from classes accelerated like a small snow slide of its own, abetted by a lifestyle he had adopted at the Forklift. Working as a dishwasher until closing time, he would clean the restaurant until 2:00 a.m., and then find a spot to curl up on the floor, a blanket draped over his body.

For Gordon, his education on the snow was headed toward summa cum laude status. Waking early, he would snag first chair at Snowbird and spend long days in the mountains, getting to know their every tendency, polishing his ski prowess, and sensing the first inklings of his interest in the inner mysteries of snow. In the winter of 1985 he was hired onto the ski patrol at Brighton ski resort. He had left college behind, a victim his parents concluded of being "Craig, the free spirit." In reality, he was anything but.

Among the "old guard" ski patrollers at Brighton, Gordon initially embraced their lifestyle. But the grizzled veterans did not embrace him. A skinny kid from New Jersey, he could certainly ski, but he had no big mountain credentials or credibility earned through deep mountaineering experience. Gordon felt that the old-timers were just waiting for him to fail, to reveal some flaw of character or capability, to quietly turn tail and retreat back to the East Coast.

Subject to the old-timers' skepticism, Gordon forced himself to expand his skills in morning avalanche mitigation work—with no willing mentor in sight. He threw avalanche release bombs into deep snowpack, to learn about the detonations' effect on layers of storm snow. He learned which bombs caused slides to release and which slides received the bomb like an explosive excavating a hole. Without any senior ski patrollers to teach him, Gordon might just as easily have blown himself up, an outcome some of the old guard might have shaken their heads over while muttering "told you so."

But something began to happen: hard work. In retrospect, Gordon sees the effect his initiative had, if not on some of the stubborn ski patrollers, at least on the management at Brighton. That first summer he ran a trail crew clearing trees and rocks from ski runs and implementing other repairs and improvements. There were rumors of Brighton developing an ambitious expansion plan, something Gordon could see himself being part of. The next winter, he worked hours beyond his usual tasks to perform "snow studies" in the resort.

While not a full-time student, he had circled back to the University of Utah and enrolled in a class called "Avalanche Forecasting and Snow Dynamics," taught by the legendary Peter Lev. Gordon was nearly hypnotized by the lectures. In mountain-climbing circles, Lev was well known as a co-owner of Wyoming's Exum Mountain Guides, one of the world's most elite guiding services. "The first five minutes of the first class, he talked about his résumé," Gordon recalls. "He was actually a pretty famous mountaineer. The light bulb went off for me: 'You can make a living on the subject of snow.' He truly changed my life."

Gordon heard things that expanded his view of what a profession in the skiing world might look like. Lev mentioned having worked with a company called Mike Wiegele Heli-Skiing, a world-renowned operation in British Columbia. Having served his apprenticeship as a dishwasher sleeping on the cold floor of a restaurant, and fending off indifference from the senior ski patrollers, Gordon saw his future anew. Suddenly the climb up the roster at Brighton that he'd envisioned as a career was just the start. The course drove Gordon to devour even more snow studies in his spare time. "I studied snow intimately. It became the fiber of my being."

Word about the potential expansion at Brighton had spread, and Gordon shifted his focus to the proposed new terrain. "At first, I thought I was going to get into trouble." But the opposite happened: not only did management encourage his snow studies and welcome the input that might shape their plans, they made Gordon, now in his fifth year at Brighton, assistant head of avalanche mitigation and the resort's snow safety expert.

With his newly earned status, Gordon was offered an even greater reward—employee housing. His education in snow took on a kind of graduate-level intensity. He'd spend his days working at it, and when big storms would roar into the resort in the middle of the night, he'd get up, don some gear, and head outside into the maw of it. "It was a very romantic time," but just as important, he was expanding his knowledge and ability to read snowpack and mountain terrain.

At the ten-year mark of Gordon's employment at Brighton, the rumored expansion had not yet happened, leaving Gordon feeling "a little stifled." But he had caught the eye of a Texas girl who had come to the resort to snowboard and ended up selling lift tickets so she could stay. "I was what you call a 'window bimbo,'" Anita Gordon recalls. "We went out on a date in March 1994 and we've been together ever since."

After winning over Anita, Gordon looked for his next move. He had gotten wind of a heli-ski operation being established at Wolf Mountain—now Park City's The Canyons. He knew that essential to such an operation would be a "snow guy," someone to forecast conditions for skiers and to identify avalanche threats, particularly on helicopter-accessed peaks where there was no bomb control except from possibly the heli itself. In December 1995 he joined Wolf Mountain Heli-Ski company as forecaster and guide. Because the elite ethos of heli-skiing is to take clients into untracked ski terrain that no one else can access, Gordon decided he would open up runs where no other service was likely to operate.

An hour drive to the east of Park City sat the Uinta Range, which for years had occupied a larger-than-life place in his imagination. "All I knew is that the story of the Uintas was that it had a mystique about being very dangerous snow. I had heard very spooky things about it. People said no one skied the Uintas because it was so sketchy." The mystique—and cracking it—was all Gordon needed. He flew reconnaissance missions into the fabled mountains, where he saw "remarkable terrain, big open bowls, and high dramatic peaks." The Uintas have a different look to them. They don't share a physical lineage with the Wasatch Range, the craggy, dramatic mountains that bear more resemblance to Idaho's Sawtooths. Surrounding Park City and backdrop to Salt Lake, the Wasatch are more like celebrity mountains, the main act.

But the Uintas have a certain visual purity. While not that geographically remote from developed areas like Park City, the Uintas feel like a true wilderness. Punctuated with steeple-like peaks that, when covered in snow, evoke a crazy pastry chef's creations, the range is whipped up with fine pinnacles. Generous snow bowls fan out in inviting, grandiose gradients similar to opera houses and concert halls, comparisons that belie their steepness.

Initially Gordon did not so much worship the Uintas as attack them. Scouting one peak, he leaned out a helicopter door and dropped an avalanche bomb into thick snowpack. "We place some explosives on the slope and it triggers several very large avalanches. I'm totally in my element as an avalanche hunter." He found a rush in hopping aboard a Bell Jetranger helicopter ready to alight with pilots, clients, and the anticipation of high adventure. But two things worried Gordon. The first was his own conduct in scouting out runs. "I was hunting rather than avoiding or forecasting for avalanches. I was on an avalanche safari." The second thing was the acceleration of growth in the business and the logistical complexities that came with it. To stimulate more business, an owner, who knew little about snowpack, once overruled Gordon's call about safe terrain. With all that pressure, Gordon sensed mounting risk, even a feeling of being part of a reckless endeavor.

IT CAME TO A HEAD in the winter of 1997–98. Gordon was guiding on a day in "big terrain enveloped in a melt/freeze corn (snow) cycle." There was concern over the stability of snow conditions coupled with the size of the client group running into the mid-teens. On top of that, they were mostly Germans who spoke no English. "We had a challenge in communication and group management." In the midst of the delicate ballet of helicopters taking off and landing, and managing clients at the landing zone while others were out on the steep slopes, Gordon feared he might be dancing along a thin edge of potential disaster.

As the afternoon tension rose, he locked on a thought. "We're either going to be going to the bar and saying, 'What a great day!' Or we're going to go to the bar saying, 'What the fuck are we doing?!' I realized I better get out of there and not have my name associated with something on the five o'clock news." His departure from heli-ski guiding led to a period even Gordon has a hard time articulating. In the aftermath, he essentially became a solo ski nomad wandering the mountains and backcountry in the winter of 1998–99, honing his mountaineering and snow skills in ways that only such freedom permitted. Although he doesn't exactly equate the two, Gordon describes this era as infused with the same sense of romance he experienced while living at Brighton and ranging out at midnight to stand in the teeth of an icy wind and blizzarding, horizontal snow. It's Thoreau in the mountains, only without the journal.

Scratching together a living in the summer doing heavy equipment work, Gordon's job opportunities multiplied as Utah and Salt Lake began a walk-up to the 2002 Winter Olympics. There had been a legacy to his winter of immersion in the out-of-bounds: on the website of the Utah Avalanche Center, he had posted scores of avalanche reports and threats based on his firsthand observations. In late 1999 the avalanche center offered him a job as its educator for snowmobilers. At first it didn't seem the perfect fit—he was a skier, not a driver of a smoke-belching piece of machinery that raced by with the wailing of a banshee. But Gordon understood he would "be affiliated with an organization I had a very high regard for." Like Peter Lev, whom he'd studied with, there were mountaineering legends involved with the avalanche center as forecasters and educators. Gordon could use the skillset he'd built and become one of them. "It was great. They saw a better fit for me than I did for myself."

Because no snowmobiler avalanche awareness program existed, Gordon created one. He soon found a different and intriguing challenge. His constituency wasn't from Salt Lake City or its suburban environs; it was largely from more distant counties where people lived a far more rural lifestyle. They had not had a graduated snowmobile experience that led them from a resort into the backcountry. As snowmobiles had evolved, from working machines on ranches to weekend recreational vehicles, there had been no structured rules or outdoor curriculum.

On workdays, throttles ran wild on the ranch, and on weekends they opened even wider, too often governed as much by alcohol as ardor. Into this, Gordon waded with his hip demeanor and elite skier's résumé. Several dynamics played out in Gordon's favor. He worked on meeting snowmobilers in their own backyards, at snowmobile dealer shops and other places. He saw the need to expand avalanche center forecasting to areas that snowmobilers frequented, including one large "black hole" begging for daily advisories. And then two tragedies led to "an eclipsing moment."

IT WAS DECEMBER 2003. The day after the Christmas holiday, up at actor Robert Redford's Sundance ski resort, above Provo, Utah, stiff winds began to blow through the ski lifts. Sundance sits in the shadow of Mount Timpanogos, at 11,752 feet. Beyond majestic, the peak demands reverence and a wary respect from those who hike its flanks. With the winds increasing, the resort made a decision to shut down the ski lifts. As they disembarked from the lifts, several groups of skiers, snowboarders, and others—fourteen in all—decided to hike into the steep snowfields toward the foot of Timpanogos.

To say the snow was deep that day would have been to state the obvious. The Salt Lake area had just experienced its second largest snowstorm in history, with five to six feet of new snow in the mountains. Wearing his forecaster hat, Gordon later called the situation the fourteen people hiked into a "terrain trap," a canyon or flattening runout below steep slopes from which naturally triggered avalanches can easily fall, pick up momentum, and bury people. The runout they had entered was dubbed one of the largest avalanche paths in Utah, prime for a massive slide. As the group of riders forged deeper into the canyon, that's what happened.

A dozen riders were trapped in the slides which, when later measured, covered an area the size of twenty-two football fields and ran as deep as twenty-five feet. Three young snowboarders disappeared altogether. The first victim was found two days later, the day after Sunday services all over Salt Lake had memorialized the missing and presumed dead. So deeply were they buried, the remaining two boarders' bodies were not found until that spring's snowmelt.

The second tragedy occurred the following winter of 2004–05. Heavy snows had begun to fall uncharacteristically early, even for Utah. By Halloween, recalls Gordon, there were some backcountry spots where the snows "were over your head." In late November, over Thanksgiving, a clear and cold stretch of weather hit, which froze lower layers of snow to the consistency of icy skid-plates. Another large storm hit in early December that "put a thick slab on top of weak snow." Within two days, four people had been killed by avalanches—a skier, a snowmobiler, and two snowshoers.

The deaths were not so much another adventure story gone wrong. Rather, in the wake of the previous winter's losses, they reverberated like family tragedies in the wider Salt Lake community, gaining attention in small, rural neighborhoods, in public gathering spots, and in churches. The dead weren't strangers; they were well known as immediate or distant family members, fellow workers and students, and social acquaintances. Gordon attributes some of the emotional impact to the ubiquitous presence of the Mormon church, whose members knew the dead and their families.

With echoes of the previous winter's accident, there was the grim aspect of the staggered recovery of the dead. It kept the tragedy front of mind, as if the Salt Lake community was quietly grieving for months. The events and fallout got the attention of the recreational snowmobilers, and Gordon found his program more welcome than ever. He attributes much of the impetus behind the development and launch of the "Know Before You Go" schools program to what happened in these back-to-back winter accidents. Gordon initially pitched the "Know Before You Go" project as one that would be experienced in its first season by five thousand kids. "At the end of that season, we had talked to twelve thousand kids. We had really eclipsed the five thousand we thought we would do. It was a defining moment in avalanche education and outreach." The video that Gordon showed in the Ogden seminar the day after Jeremy Jones's accident in the Uintas has been released in multiple languages and seen by hundreds of thousands of people.

Today, Gordon is one of eight forecasters at the Utah Avalanche Center, which is overseen by a director, Mark Staples. While car accidents, rogue storms, hurricanes, and tornados kill far more people than avalanches, the impulse among young people to escape into the backcountry grows every year. The need for the sometimes brutal truth of what can happen when threats aren't taken seriously is greater than ever. While Staples has witnessed this dramatic jump in backcountry snow sports, he says the core mission of the center hasn't changed. Technology—web postings, social media, and a podcast—has enhanced the content and speed of avalanche advisories. But retaining a highly personal "voice" (dating back to pre-internet, daily phone reports) is essential. "We want people to know there's a human behind every forecast," he says. "It's about connecting with people on an emotional level." With such connections, Staples emphasizes, comes trust. "We always will have to maintain our credibility. That's our biggest asset. . . . Keeping a human face behind that is important. . . . Our forecasters can't just be snow scientists. They have to be communication experts."

Forecasters like Gordon spend more time than ever in the backcountry because they can post updates remotely and instantaneously, meaning more timely pinpointing of rebellious snowpack. Gordon's own constant exposure to avalanche dangers has made him think more deeply about risk. When asked if he ever fears dying in an avalanche, Gordon resorts to humor and a touch of pathos. "I would die of embarrassment before the avalanche would even kill me," he jokes and then grows solemn. "I have made such a connection with people throughout this valley, people who don't even have anything to do with snow or skiing, and I feel like I would disappoint them so greatly. . . . I feel like if something happened I would let a lot of people down. Quite frankly, I think that makes a lot of decisions for me. In a lot of regards I'm super cautious and very conservative. And maybe in some regards it's a good thing, because it gives me an easy out."

Still, the world Gordon patrols can be merciless. In one segment of the "Know Before You Go" video, a group of skiers hike to the peak of an extremely steep mountain. The first skier stands at the edge, pauses, breathes deeply, then drops in. His companions lean over to watch his first turns, of which there are four. Then the skier disappears into the massive, whitewater-like wave of an avalanche. By the time his body is located, he's dead.

Fighting back tears, one of his companions recalls the moment and says: "No one's life is worth four turns."

6

TRACKING THE DRAGON AT TEN THOUSAND FEET

Out by the Salt Lake City airport runways, in a low-slung building, Randy Graham and his team of meteorologists had been tracking for days the storm that almost killed Jeremy Jones after it blasted the California Sierra, then bore down on Utah's Wasatch and Uinta Ranges. Graham, head of NOAA's Salt Lake operation, had marveled at the snowfall, which he and his fellow forecasters usually refer to as "precipitation."

"They were getting slammed," Graham says, thinking about California's mountains. The meteorologists calculated that California had probably received four times the normal effect of a high-altitude, sustained cloudburst. Records had fallen as fast as the rain and snow itself. As the team monitored the storm working its way across the state, Graham looked at computer maps of California with red dots indicating places where records had been broken. On January 8, 2017, the red dots nearly blotted out central and northern California. Three subsequent days set records for which there were no modern precedents.

They watched the storm, swiftly traversing the Sierra at ten thousand feet and above, accelerate toward northern Utah's Wasatch and Uintas. By the morning of January 11, at the 9,300-foot summit of Park City mountain, twenty-eight inches of snow—nearly two and a half feet—had fallen in twenty-four hours. At the Canyons, another Park City peak reaching 8,800 feet, twenty-six inches was recorded. Near the base of Alta ski area, at 8,799 feet, twenty-four inches was logged. But what caught Graham's attention was the water content of the snow. Average snowfall for Utah, because it tends to be drier snow, produces one inch of water for every thirteen inches of snow. That would mean the snow totals for January 11 would yield about 2.3 inches of water. But the water totals for Park City were ranging from 2.8 to 3.6 inches; at Alta the water content was 3.5 inches. For an avalanche forecaster like Craig Gordon, this was a clear sign that the snow dragon had crept by night into the Wasatch and Uintas, and that avalanche conditions were at redline stage.

This particular storm, with its sheer scope and prodigious snowfall, embodied a lot of what had captivated Graham about weather when he was a boy growing up in Omaha, Nebraska. He recalled a time in 1975 when he was just five. That winter, a blizzard had crippled the city, and the storm's power and presence in streets, fields, and yards had left Graham not so much with a fear of nature and weather, as with a burgeoning wonder about its seeming randomness and mastery of the world.

Four months after that storm, later in the spring, an EF-3 tornado set down in Omaha. Graham once again was exposed to nature's ferocity. Within this youthful crucible, his incipient meteorologist's curiosity and investigative mind were conceived. At the University of Nebraska, he studied weather in earnest and began chasing storms "before it was a cottage industry." Out on the plains, Graham hunted down thunderstorms and hoped for glimpses of the inner secrets of tornados. In 1993 he joined the National Weather Service in Sioux Falls, South Dakota, where he apprenticed with veteran forecasters. After a few years, Graham earned the chance to apply for a forecaster job and spotted an opening in Salt Lake City. In 1995 he landed in the shadow of the Wasatch Range where, he admits, his weather learning curve was as steep as the mountains themselves.

"I didn't know a thing about mountain meteorology," Graham recalls. But he had preserved his childhood curiosity about the vagaries of different climates, and he applied it to trying to understand the often mercurial microclimates of mountains. "Weather is very complicated in mountainous terrain. The conceptual models and notions we have in our heads about weather—mountains sort of tear those apart." Particularly in America's intermountain West, Graham says, resorting to a rare bit of sciencespeak, there is "a spatial incoherency in the precipitation." That translates to a dual, seasonally driven weather problem for a state like Utah. In winter, storms like the January 11 blizzard create havoc in the mountains, closing down roads, knocking out power, and sometimes killing people in avalanches. In summer, what Graham calls a "fire triangle" develops. Fuel, in the form of dry trees and brush, combines with oxygen in the form of winds and heat, often from lightning strikes, to feed wildfires. Heavy rain hitting sun-scorched, stone-hard earth, or slopes of slick rock canyons, can create flash floods. Even the West's multistate dynamics of weather can add to the hazards. Downslope windstorms from Wyoming, for instance, can break over the Wasatch like an enormous wave of air that fells trees and may rip into homes and buildings. "It's like fluid dynamics."

Graham spent his first tour in Salt Lake City for three years, then became a senior forecaster and science operations officer in Grand Rapids, Michigan. But he missed the complexity of mountain weather, the just-around-the-corner surprises it could bring. So in 2005 he returned to Salt Lake as the NOAA region science operations officer, a job that pushes the edge, through data, technology, research, and evolving climate science, to help forecasters derive better conclusions about approaching weather. In this latter role, Graham focused on studying things like the attributes of lake-effect snow produced by the Great Salt Lake, how to better predict levels of rainfall in flash floods, and the origins of downslope windstorms. In 2015 he became the meteorologist in charge of NOAA's Salt Lake office.

Handling storms like the one that hit on January 11 has evolved, Graham says, into a much wider collaboration than it used to be. Forecasters may have superior, even proprietary data on weather systems, but if it's not shared efficiently with avalanche forecasters, law enforcement, and municipal and state officials, it falls short of its potential uses. "One of the things that led us to where we are now was the Winter Olympics of 2002," Graham explains. As part of the multifunctional team overseeing preparations for the games, he got exposed to all manner of disciplines. When the planning began in 2001, it was just after the terror attacks of 9/11, so Graham was on a team that included extensive blueprints for security, which extended to extreme weather events. But there were also fine points that had to be explored. At Snowbasin ski resort, where the Olympic men's and women's downhill races were to be held, a private meteorologist was employed to complete a climatological study so the race organizers could understand what level of crosswinds might affect the racers. Another weather scientist evaluated what winds would be like under different weather conditions at the ski jumping ramp in Park City.

For all his passion about work with the Olympics committee and snow professionals at ski resorts, or Craig Gordon and his fellow forecasters at the Utah Avalanche Center, Graham's earliest enthusiasms about weather remain his inspiration for exploring weather phenomena. "All weather is the atmosphere trying to bring things into balance," he says. "The atmosphere is always trying to be stable, to find equilibrium." Because weather is moving, growing cold and hot, wet and dry, and aroused by terrain variances—flat land, deserts, mountains, lakes, rivers, and oceans—storms spring up trying to close the proverbial gaps between these discrepancies.

All of which takes Graham back to the chain of atmospheric rivers of early January and the storm dragon of January 11. Climate scientists, he says, have noticed what they believe is a change in how storm patterns embed themselves in geographic regions. Historically, a period of very hot or cold weather might yield to moderation, or a lone, massive winter storm might roll across the Sierra into the Wasatch. But unusually prolonged spells of intensely hot or cold weather, or large winter storms roaring one after another, or home-flooding rains over a single city—these repeated, major, contiguous events have caught weather scientists' attention. "One of the theories in climate change is that we're going to get stuck into patterns longer. That, in theory, could explain what we've experienced the last two winters. I can't say it's definitely the case. But it's a pattern we've noticed."

Senior hydrologist Brian McInerney is a Salt Lake weather service colleague of Graham's who serves as a regional lead on climate change. "We're getting hotter all the time," he says. "Utah is warming faster than the global average. Utah is warming at two to three degrees above what is considered the average from about 1950. It's in a desert realm and that's mainly the reason." In addition to the weather patterns Graham notes, McInerney has tracked a trend of variable winter weather—some winters drier with less snow, and others wetter and warmer, which means rain in the high mountains, even at ten thousand feet. All this can create unstable snowpacks, which can lead to avalanches. "When you think of a healthy snowpack, you get snow early and you get snow often," McInerney says. "The layers don't have time to turn into faceted snow, so they bond really well. What you don't want is long periods of time between each storm, or to have a weak layer underneath when the winter started late." When those storms come through wetter than historical norms, it "means more water content, and heavier snow. Avalanche activity picks up in a big way."

While Graham, McInerney, and other NOAA scientists are primarily focused on weather events that affect the United States, climate scientists around the globe have studied how rising temperatures may change the behavior of snowpack in the mountains. Early conclusions are leading them to a prediction—more avalanches. A team of researchers from the University of Geneva, working with officials in India, formed the Indian Himalayas Climate Adaptation Program to determine how changes in snowpack might affect the more than forty million people who live in the Himalayas. They found, in a March 2018 report, that warmer air in the Himalayas—the so-called Third Pole outside of the Arctic and Antarctic—is increasing water content in the snow, destabilizing it, and creating more avalanches.

"Avalanches are bigger, travel greater distances, and are triggered earlier in the year," the researchers found. "And rising air temperatures are also affecting the cryosphere: glaciers are receding and permafrost is melting, losing its role as a sediment stabilizer." They point to a 2014 accident on Mount Everest when, early in the climbing season, sixteen Sherpas were killed by a huge ice avalanche spilling down from the flanks of the world's highest mountain. A year later, twenty-two people were killed in Nepal when an earthquake struck, triggering avalanches from unstable snowpack. And in late September 2019, Italian officials shut down roads and forced home evacuations in the Alps for fear that 250,000 cubic meters of glacier ice on Mont Blanc was showing "significantly increased" signs of cascading down into occupied areas. Some nearby Swiss monitors were even holding a "mourning ceremony" over the disappearing glaciers.

A decade ago, Graham says, some people would point to a specific weather event and try to use it as evidence of climate change. But disciplined science and experienced researchers like Graham knew that back then the tools to correlate a single event to longer-term changes in climate were rudimentary at best. Today, thanks to supercomputers, weather scientists are "fingerprinting" weather events, particularly events in close succession to see what formation and effect traits they share. Graham explains: "In meteorology, we understand the physics pretty well. In a general sense, weather is driven by Newtonian physics." However, he readily admits there's a vast amount of weather behavior he and his colleagues don't understand. Even the current NOAA forecast models allow for a measure of guesswork beyond several days out. "As you go out in time," Graham says, "errors begin to expand for a variety of reasons: imperfections in the starting point of the model, important interactions happening at scales the model cannot capture, or phenomena that aren't completely understood."

He recalls a US Air Force forecaster he once talked with who summed up a kind of eternal forecaster's dilemma.

"Randy, you know where the toughest place to forecast is?"

"No," Graham replied.

"Wherever you're at."

Graham thinks about the snowboarders who were caught in the Uinta avalanche and the storm dragon that swept them down the mountain. He sees the work he and his fellow forecasters do as supplying scientific data to the public so they can use it to plan their daily lives. Sometimes experts like Craig Gordon have to refine that data for specific purposes like venturing into the backcountry. "Those guys are rock stars," Graham says.

Even with all the forecasting expertise, Graham has seen his share of predictions fall prey to massive, unruly snow storms. "Storms do have character," he says. "Some just wreak havoc."

7

ALCHEMISTS OF THE SNOW: THE FORECASTERS

In the days after investigating the avalanche accident in the Uinta Range, Craig Gordon was especially sensitive to the hair-trigger, perilous nature of the snowpack he was dealing with. A day or two wasn't likely to alter the snow structure. More than two feet of newer heavy and dense snow sat on top of older layers, decomposing like rot eaten through by wood bores. After seventeen years it didn't feel like one of his finest moments. Despite what had happened to Jeremy Jones and his crew, Gordon felt he'd done his job well. His warning had leveraged the red-flag words—"avalanche danger is high." His forecast had been right on the money.

But he knew that when you sign up to guide other people into extreme environments, you had to try to protect them not just from objective threats but also from themselves. So he kept at it. The day after the accident he posted an advisory that read "the avalanche danger is HIGH today. Very dangerous, human-triggered avalanches are CERTAIN on steep, wind drifted slopes.... These are the kind of snowpack conditions that kill most all-mountain riders."

The day after that, Gordon's posting reiterated this initial warning, word for word. Then he added more detail, revealing how his avalanche forecaster's mind was sifting through the evidence from the investigation. "Once triggered, avalanches are breaking deep and wide.... snapping trees along the way. They're definitely something you don't want to be on the receiving end of!" Focusing specifically on the accident, Gordon noted: "What we do know is two snowboarders are very lucky to be alive after triggering this four feet deep by three hundred feet wide avalanche which broke to weak, faceted snow formed in late December."

Now Gordon was moving into the forensics of what he'd seen at the site of the avalanche—the telltale decaying layer of snow from a December storm. He added his final appeal to backcountry riders to try to imagine what he had seen: "The problem is complicated because the snow will feel strong and bomber under our skis, board, or sled. But here's the deal, we've gotta think not only about the snow we're riding in, but also the snow we're riding on and there's a few buried weak layers that are straining to adjust to all this added weight."

This is the daily equation Gordon and his colleagues at the avalanche center have to contend with. Initially they're presenting backcountry riders with hard information—snowfall totals, wind speeds, temperatures, the variations between north- and south-facing slopes. But at some point, there's an emotional dimension, a bid to get riders to think the way they do, to visualize the snowpack as if the riders were snow geologists examining a thick column of snow and ice, some of it solidified, and some as granular as sugar in one's palm.

Avalanche forecasters try to temper the impulses of inherently impetuous riders. Skiers see a fresh foot of snowfall on their front lawns and they want to rush into the backcountry at sunrise. It's the seduction of "first tracks." "Once everything is white," Gordon says, "it all looks the same." At least it may to some riders. Ultimately, the forecaster's job is a bit of a tango with the devil's dragons. In the dark morning hours, deciphering all the data from sensors in the mountains, avalanche forecasters know it will ineluctably lead to a conclusion that must be shared with people who trust the forecasters with their lives. They might ignore the warnings. But even when they do, and fall victim to an avalanche, there's a certain assignment of collateral blame to the forecaster. Was the forecaster's warning strong enough? Could he or she have tapped into more information on the snowpack? Was he or she too tired at 4:00 a.m. to recall the layer of *graupel* snow, tricky as ball bearings, that fell five weeks ago? Or was the victim a hapless, unwary thrillseeker sent on a coffin ride down the mountain because a forecaster was reluctant to cry wolf too loudly? It's a complex intersection of snow science, judgment, and human urges.

"What surprises me about snow," Gordon says, "is how much we get away with, how many close calls there are." As if to put a finer point on the fluffy illusion snow presents to some riders, and the tendency of many to try to take advantage of that, he explains: "Snow reinforces a lot of bad behavior."

There's an intellectual echelon of backcountry riders who can be duped by the lure of untracked snow that, to their eye, bears no evident sign of avalanching. Bruce Tremper, who spent almost thirty years as director of the Utah Avalanche Center, quotes the late Swiss scientist Andre Roch, "The avalanche doesn't know you're an expert." But the allure of fresh powder snow to backcountry skiers and boarders remains strong, and even avalanche forecasters can be beguiled by it.

IN LATE FEBRUARY 2017 a couple of perfect snowstorms breezed through the Utah mountains. On the morning of February 23, Gordon posted an advisory: "Yesterday was probably the best day of the year and on a scale of 1 to 10... it registered an 11 on the 'Sickter' scale! ... Wow ... what a right-side up, designer storm!" The next day, he came back with an even more tantalizing report that seemed aimed at prompting riders to call in sick to work: "Yesterday's cold impulse produced an additional foot of ultra-light, chin-tickling, Utah überfluff and storm totals are nearing thirty inches.... On a go-anywhere supportable base, riding and turning conditions are about as good as they get... yep, it's over-the-head and over-the-hood."

All this—the threats that lurk beneath, and the thrills that float above—speaks to the roulette wheel that forecasters as well as backcountry skiers and boarders spin on any given day. You can plan the best route, carry the newest beacon and probe pole, digest the most detailed avalanche advisory, and still your number can come up in the devil's lottery. Or, as Gordon characterizes it, in the snow dragon's lair. And that's a place he strives to avoid.

"It's one thing early in your career to be curious and to test theories" in the backcountry. "It's another thing to think you are going to outsmart Mother Nature. Man, you are fooling yourself if you think that's the case. Because if you are going into this game and that's the hand of cards you are playing, man, it's only a matter of time."

THE JOB OF AVALANCHE FORECASTER is not exactly something one aspires to when undergoing career counseling at the local high school. People don't really prepare for it in university curriculums, although increasingly a striking number of schools have degrees in meteorology, geology, or engineering coursework that prepares students for better assessing winter storms, studying layers of snow as if they were rock striations, or for understanding at what point structural weight and pressures cause things like bridges or snow cornices to suddenly come crashing down.

Utah native Wendy Wagner, who became director of the Chugach National Forest Avalanche Information Center in Girdwood, Alaska, came up through the avalanche forecasting ranks in a less traditional way. Though she had an extensive skiing background, she built a base of knowledge through mountain atmospheric sciences study at the University of Utah. "My favorite class I ever took was cloud microphysics." As she listened to a lecture on how nearly invisible snow crystals form and exponentially colonize in the upper atmosphere, Wagner thought, "I don't want to be anywhere else in the world. This is amazing."

Historically, most avalanche forecasters have served extensive time and responsibilities as ski patrollers at resorts where they have become involved in avalanche mitigation work. Many, like Craig Gordon, end up taking some kind of continuing study courses in weather science. Most rely on their experience, not only in avalanche mitigation and rescues but also their acquired topological knowledge of specific mountain ranges. Being able to visualize familiar peaks, pitches, steep couloirs, and terrain trap runouts and to marry those images to new snowfall enhances avalanche forecasters' advisories. Such advice can pinpoint danger areas that might mean the difference between a day of delightful touring or death.

Wagner, now forty-four, is a former Nordic skiing Olympian, who competed in two Winter Olympics, the 2002 games in Salt Lake and the 2006 games in Torino, Italy. Although her professional skiing focus was cross-country, in the mid-1990s she became an adept backcountry skier who logged serious time in the out-of-bounds. She experienced a close call in the backcountry outside Gunnison, Colorado, when she was one of five skiers out for what seemed like a laid-back day of touring. Four of the skiers dropped into a slope and skied down with no problem. The fifth took off above them and provoked an avalanche that ballooned to lethal proportions—three feet deep, five hundred feet wide, streaking down fifteen hundred vertical feet of mountainside.

Wagner and two others managed to escape by backing up behind nearby trees; a fourth person skated to the side of the onrushing slide and "just barely" missed being caught, Wagner recalls. The skier who'd started the slide was swept below and miraculously was unhurt. Until then, Wagner's time in the backcountry had seemed, like her Nordic skiing, a joyous way of getting exercise in the midst of natural beauty. "That was my first exposure to the understanding that this could not be a good situation." Still, it was five years before she took an avalanche awareness course, which started to fill in the gaps for her related to reading backcountry terrain and spotting avalanche danger.

In Alaska, Wagner's work involved an all-woman staff of three. In addition to her academic study, Wagner got involved with the Utah Avalanche Center and received extensive mentoring from the staff. "The reason I'm able to have this job is because they took me under their wing." After applying for a forecaster job in Alaska, she got a phone call in January 2011, while touring in the backcountry near Alta. Uncharacteristically she took the call because it registered on her phone as an Alaska number. Normally it would have been bad form, she explains, to take a phone call in front of fellow touring partners while on the snow.

"Two weeks later, I was in Alaska and I've been here ever since."

LIKE GORDON AND MANY OTHER forecasters, Doug Chabot acquired his avalanche forecaster's résumé through ski patrol work as well as extensive rock climbing. Today, Chabot is director of the Gallatin National Forest Avalanche Center, where he's managed the operation since 2000. With a team of two forecasters, Chabot is entrusted with overseeing a vast six-thousand-square-mile swath of Montana mountain ranges.

After attending Prescott College in Arizona, where he fell in love with climbing and began to learn about avalanches, Chabot took an internship in Silverton, Colorado, where much of the skiing was double-black-diamond difficult. "The hardest part of the day was getting down," he recalls. Chabot moved to Bozeman, Montana, in the winter of 1986–87. Working at a Big Sky ski area restaurant called Whiskey Jack's, in his free time Chabot struggled to master the basics of skiing. During his second Montana winter, he became a bartender at Jimmy B's, a Bridger Bowl ski area watering hole. When he wasn't serving up cocktails, he pestered the ski patrol for a job.

The problem, Chabot says, was that when the patrol evaluated his skills, they told him he was at best a haphazard skier. Still, he continued to badger them. Finally relenting, the ski patrol essentially required him to moonlight on his job. "I think I was the only skier on patrol who *ever* took ski lessons." Chabot simultaneously honed his skills and took on more challenges in his climbing. Now that he was working as a snow professional in "big mountain" terrain, he shifted his climbing from sport style, usually several-hour ascents, to extended, multiday, alpine-style climbs. "Climbing for twenty-four, thirty-six hours straight, I had an affinity for that. If you have a short-term memory, and you're willing to suffer, you can be an alpinist."

In 1995, Chabot guided his fifth and last climb up Denali in Alaska. He considers that experience "a turning point" in his commitment to mountaineering and seeking greater professional accomplishment in the outdoors. He had started to make a name for himself in the western climbing community and drew the attention of renowned climber Jack Tackle, who invited Chabot to meet him in Alaska and make an attempt to scale a route called *Elevator Shaft*, a feature on 8,460-foot Mount Johnson. Tackle had tried the climb two years earlier and failed to make the summit when he and a partner became trapped as a storm savaged them. "What a death route!" proclaimed one of Chabot's friends when he told them about his plans.

Chabot, now fifty-four, has seen enough in the backcountry, including the deaths of many friends and avalanche victims, that he believes both skill and luck play a role. "If you're doing a lot in the outdoors, luck matters. . . . It comes through, and thank God it does." But he is quick to emphasize the importance of knowledge and judgment. "If you're going to have longevity, you have to know when to say no."

He and Tackle made it to the top of the route. Chabot looks back on the achievement as a youthful rite of passage that ushered in even more ambitious climbs in places like Pakistan, where he has made repeated trips. "It was pretty scary stuff," he recalls of his *Elevator Shaft* summit. "It was good luck, it was good weather."

ONE THING WAGNER SHARES WITH other forecasters and backcountry professionals is an almost magnetic pull toward wanting to explore and understand how snow looks and behaves in all its different incarnations. Listening to avalanche forecasters talk about snow, it's like they have acquired an intimate, encyclopedic mental file cabinet of extensive snow images. To the unpracticed eye, snow is snow. To a forecaster, snow is victim and perpetrator, wealth and poverty, unrepentant wallflower and the drunken, loud-shirted miscreant of the party.

"In my dream world," Wagner says, "I wanted to understand mountain weather, but also to study snow on the ground." Craig Gordon echoes this feeling as he describes living at Brighton ski resort and dashing out, in the middle of the night, to confront blizzards so he could better understand the differences in storms and how snow accumulates and what characteristics it displays.

Thirty-one-year-old Shannon Finch also seeks to understand the snow. After ten years as a ski patroller at Sundance ski resort, she now works as a backcountry guide with Park City Powder Cats, a snowcat operation in the Uinta Range. When she was still building her skillset, Finch often ventured into the backcountry by herself on her days off. One of her preferred ski tours was in an area called Aspen Grove, along the north-facing side of Mount Timpanogos, the majestic 11,752-foot peak that looms above Sundance.

"I'd go for a tour and look at snow," Finch says of these solo forays. "It wasn't so much about skiing as looking at snow, how it compacted compared to [snow at] the resort." She emphasized improving her route-finding abilities, a skill that resort skiing, with its color-coded and named-run signs, shapes into an experience more akin to steering down a boulevard. "I have not seen a lot of snow move in the backcountry," she says, alluding to avalanches. "I think a good backcountry skier is avoiding avalanches." Where Finch has seen snow shifting and spilling down the mountain is in-bounds at Sundance when doing avalanche mitigation work. "This has given me plenty of opportunities to see the snow move, predict its potential to avalanche and then test my theories with the use of ski cuts and explosives. This experience has become invaluable to my work in the backcountry . . . as I can better predict what conditions will create avalanches."

Chabot says his need to read snow and detect avalanche danger came from the threats he encountered early in his mountain climbing in Alaska. His first ascent was in 1989, when he was twenty-four. "You had to understand snow, you had to understand avalanches."

Some avalanche forecasters move on to higher paying jobs that leverage various aspects of their snow skills and command of mountain weather. A few gravitate to academic roles as the demand for outdoor and recreational degrees grows. But the tenure of many avalanche forecasters is strikingly long. They need the proximity of the mountains and the different stories mountains tell every day.

Wagner talks about occasions when she's riding a snowmobile in the Alaskan backcountry, checking out cinch points where avalanches often start to reveal themselves. She will come to a spot where the far northern sun is starting to decline at 2:30 or 3:00 p.m. and the light is just so on the mountain. "It's those emotional times," she says of this experience. "And then being out on a glacier and it's unbelievably beautiful. And then I think, 'Oh, my God, what a job I have.'"

"I get to be outside, which I love," Chabot says. "It's not boring, it's so dynamic. I love that. I need that in my life."

When Gordon isn't teaching avalanche courses or forecasting, he is inclined toward being a bit of a jokester and a generator of laughter. But he becomes far more wistful when asked about his feelings toward what he does: "It's a romance."

WHEN BRUCE TREMPER TALKS ABOUT his twenty-nine years as director of the Utah Avalanche Center and his long apprenticeship before that, he reaches back to the roots of American avalanche mitigation in the late 1940s. He seems compelled to summon the names of two men whose work underpinned his own—Ed LaChappelle and Montgomery Atwater.

Atwater arrived in Utah's Little Cottonwood Canyon after World War II and almost immediately was exposed to the canyon's seismic-like avalanches. Tremper paints an image of Atwater as a romantic crusader, a writer of books with such titles as *The Avalanche Hunters* and *Snow Rangers of the Andes*, a man whose journey defined him as an Ahab of the snow. "He started out," Tremper explains, "to tame the Great White Dragon."

Yet Tremper, through his own three decades at the avalanche center and his authorship of definitive texts on the backcountry and avalanche avoidance, assumes a bit of Ishmael's chronicler role. Tremper is also a teller of the tales that form the lore around Little Cottonwood Canyon as it evolved into "the laboratory" for avalanche learning and attempted control of one of nature's fiercest expressions. Atwater didn't die in an avalanche but rather of a heart attack in 1976 at the age of seventy-two. And at sixty-six, Tremper remains beloved in the backcountry community for his work carefully crafting the expertise and prestige of the Utah Avalanche Center.

Under Tremper's guidance the center developed training programs and an operational model for funding and community connections that became the envy of other American avalanche centers. His books—*Staying Alive in Avalanche Terrain* and *Avalanche Essentials: A Step-by-Step System for Safety and Survival*, among others—are widely read by novice backcountry travelers and by hard-core free skiers looking to fortify their existing trove of knowledge with reminders of how simple mistakes can quickly become irreversible tragedies. Tremper, who bears a passing resemblance to the actor Ed Harris, a small gap between his upper front teeth, reminisces about the early days of avalanche study, but he doesn't come across as a man living in the past. He's revising one of his books for an updated third edition, and if his physique is any indicator, he's a frequent habitué of the backcountry.

Growing up in Missoula, Montana, Tremper started skiing shortly after he learned to walk. His father, a petroleum distributor, grew up on skis and was one of the earliest volunteer ski patrollers in Montana. He was also a University of Montana team ski racer. In 1964, Tremper's father attended one of the first college avalanche courses taught by Dr. John Montagne at Montana State University, after which he taught his ten-year-old son the basics of avalanche awareness. Eventually Tremper's mother, a lawyer and university professor, took a position at Montana State in Bozeman, where Tremper ultimately earned a master's degree in geology, specializing in avalanche science and studying, as his father had, under Montagne.

After a successful college ski racing career, Tremper went to work on the Bridger Bowl ski patrol, using his avalanche knowledge to control in-bounds threats. The other ski patrollers, Tremper recalls, would warn him: "'If you go out-of-bounds there are avalanches out there and you are going to die.'" But his life changed when he met Duain Bowles, an avalanche researcher at Montana State. Bowles regularly skied at Bridger Bowl and often headed out-of-bounds to dig snow profiles to test stability. Once done digging several snow pits to test the stability of the snow and pronounce it safe, Bowles and his partner for the day would ski fabulous, untracked snow to the bottom of the "forbidden" terrain. "Wow," Tremper thought, watching. "I want to learn how to do that." He quickly made friends with Bowles and, in his words, "followed him around like a puppy dog for several years."

In 1980, Bowles founded the Utah Avalanche Center and Tremper began to make trips there. "He skied lots of backcountry powder and had a free season pass to all the ski areas," Tremper recalls. "I thought, 'This is the job I want.'" In those days, Tremper says, it became the rage for avid skiers to flirt with the backcountry using telemark skis and climbing skins because it was a European tradition. Skiing, as a mainstream sport, had come to America initially during the 1940s, after the war ended. It was embraced as a recreational endeavor and people, particularly in the American West, wanted to explore skiing's European roots. Knowledge of backcountry techniques was limited at the time, and the Utah Avalanche Center was created to provide critical avalanche information and teach awareness classes to those who wanted to venture into the backcountry.

Tremper landed his "dream job" as director of the Utah Avalanche Center in 1986. Reflecting on his involvement with the center's history, he tells a story about its beginnings with a single, rotary dial telephone at Alta ski area on which had been recorded a daily loop message on avalanche threats. A single backcountry skier at a time—there were no snowboarders or snowmobilers then—could call in. If the person got a busy signal, they'd have to call back. Eventually, as more people began going outside resort boundaries, the center expanded to three phones. As the audience rapidly grew, a bank of digital voice messages carried the avalanche updates over twenty phone lines; in the mid-1990s, all advisories moved to the internet.

Tremper gives real credit to the early US snow pioneers, Atwater and LaChappelle. In many ways they were the odd couple of the snow professional world. Atwater was a Harvard-educated liberal arts major, who had served in the 10th Mountain Division in World War II, training soldiers in winter warfare. He'd been wounded during the war, but in at least one old photo he strikes a pose at the edge of a snowy cliff as if a Hemingway on skis—robust, ready to conquer the slopes and the literary world.

By contrast, LaChappelle was a scientist, an expert in the arcana of glaciology. As Tremper explains: "Monty Atwater was not a scientist, he was an English major. He needed a scientist and Ed LaChappelle was the perfect person." Although Tremper never met Atwater, he lived for four years at Alta in a small Forest Service cabin built in the early 1950s to house Atwater, LaChapelle, and others. When Tremper conferred with LaChappelle on snow science matters, it was as if the man was inscribing elaborate equations on a chalkboard.

As an engaged student at LaChappelle's knee, Tremper was attentive, but he realized early on that a critical part of his job was to effectively communicate avalanche dangers to the public. He knew doing so in scientific jargon wasn't going to captivate anyone. "Ed would say to me, 'Any rapid change in the thermal or mechanical energy state of the snowpack is a precursor to avalanching.' Hmm, I thought, that's too complicated. So in my lectures I'd use simplified language: 'Snow does not like rapid change.' It's a phrase that is repeated like a mantra in modern avalanche classes. But the idea originally came from LaChappelle."

Fortunately for Tremper, he learned the theoretic foundations of snow mechanics from LaChappelle, Montagne, and Bowles. Bridger Bowl ski patroller Doug Richmond, who had a masters in geology and was a childhood ski-racing friend, was another important mentor. Inspiration for Tremper's communication style, however, came from two mentors in Alaska. Prior to becoming the Utah Avalanche Forecast Center director in 1986, Tremper spent a year at the Alaska Avalanche Forecast Center working with Jill Fredston, who also taught at the Alaska Avalanche School. Founded in 1976 by Doug Fesler, the school was one of the first in the US dedicated to teaching avalanche safety. Fredston and Fesler had developed a highly interactive curriculum for avalanche awareness. "As a team, the two of them were just amazing," Tremper recalls. They had built wooden "tilt boards" on which they heaped flour and sugar to simulate different layers and textures of snow. As they demonstrated increasing slope angles by slowly tilting the boards, the granular sugar would inevitably crumble and ferry the flour down with it. Cardboard boxes represented snow slabs, and small paper cups stood for weak layers of snowpack. The ersatz snow wiped away everything in its path, and avalanche class attendees couldn't get enough of it.

WHEN TREMPER ARRIVED BACK IN Utah to begin work at the avalanche center, he went to one of the center's classes and sat in back to monitor the educator's content and style. It was all drawings on flip charts and dry classroom pedagogy. "I sat in the back thinking, 'Oh, my God, this has got to change.'" So he radically altered the class and took his new avalanche show on the road. Hauling along a big box filled with tilt boards, Tremper revealed his magic with sugar, flour, cardboard, and paper cups. Copying something he'd seen at the Alaska Avalanche Forecast Center, he used Silly Putty to illustrate the viscoelastic nature of snow and rubber bands for visualizing snow's stored elastic energy. Models of Plexiglas and foam demonstrated how a load of new snow creates shear forces in snowpack. "Everything was very interactive. People would flock to those classes," Tremper recalls.

He also encouraged his staff to have some fun with their telephonic forecasts, injecting quips and first-person details into them. "We tried to make them as entertaining as possible. One of our forecasters, Brad Meiklejohn, used the phrase 'encyclopedia snow,' meaning it was so heavy and sticky it felt like someone loaded encyclopedias on the backs of your skis." Meiklejohn also came up with a tagline for the avalanche center: "We help keep people on top of the Greatest Snow on Earth, instead of being buried beneath it." Today's shortened tagline—"Keeping You On Top"—is inspired by the original mission statement.

Succeeding at one of the center's core missions—education—was gratifying to Tremper as he led the organization through the 1990s to 2015. But as a US Forest Service employee, overseeing an operation largely funded by the federal government, he had to reinvent the budgetary wheel every few years. The center needed more financial stability, and it came in the form of a ski swap sale to raise money privately. That idea grew into the formal nonprofit arm of the center, Friends of the Utah Avalanche Center. An annual fundraiser in September brings out the Salt Lake City backcountry community and adds thousands of dollars to support the budget; additional fundraising from outdoor gear companies and private individuals has led to most of the center's funding being private. Over time, other US avalanche centers have adopted the hybrid funding model with similar success.

But like anything new involving raising money, there were times it seemed touch and go to Tremper. "Looking back on it, I'm amazed we made it all this way," he says. Sitting in the living room of his Salt Lake City home, Tremper is relaxed, his sock-clad feet propped up on a coffee table as if he's inclined to take advantage of any moment when he doesn't have to wear ski boots.

Of his former boss, Gordon says: "I would give praise to Bruce because, in the early days of the Utah Avalanche Center, partnerships were established and information conduits. . . . That information highway is a real important piece of the puzzle in how we work." The center's influence under Tremper grew exponentially and its programs were embraced by the wider Salt Lake community. Of this period, Gordon says: "We were creating rock and roll." Tremper has seen the center's influence increase through online, almost real-time backcountry reports, social media, and expanded avalanche awareness classes. He's also witnessed the effects of more people venturing into the out-of-bounds.

"Just from going out there and looking at all the tracks," Tremper says, "you just go, 'Wow!'" Because many of these new backcountry adventurers are young, Tremper gets questions about how to get a job like his, where a day or two—or three or four—a week in the backcountry is part of being in the office. "You have to do this out of love," he tells aspirants. "Not because you're going to make any money out of it."

THRUSTING YOURSELF INTO TERRAIN CALLED "the out-of-bounds" is to a large degree an examined decision to go into hostile country. It's like a free diver plummeting a hundred feet or more into the sea and a world without air. The backcountry is a world without absolute safety reference points. Just white, cold land, without readily discerned visible boundaries except those formed by mountains with thousands of feet of sheer, jagged rock down which you can dive-bomb to your grave.

Out there, there are no catchy ski run signs. There are no ski lifts swinging playfully overhead, no patrol people in red jackets bearing white crosses that convey the message that medical help, should it be needed, is close at hand. So your mindset shifts in the backcountry. If you are inclined to paranoia, there's a presence all around of a latent, shimmering violence in the snow. The farther you venture away from civilization, the more you are waltzing with snowpacks that may be harboring deadly tendencies.

On many runs within ski resorts, the snow has been groomed, a little bit tranquilized, even if it is still capable of allowing skiers and snowboarders liberal bouts of recklessness. In the backcountry, snow can be out of control, emerging in the morning sun from a night of temperature- and stress-induced meteorological drunkenness. All that constrained disorder is just waiting for a trigger. To put it most simply, snowpack has difficulty adjusting to rapid change.

Though interest in backcountry skiing is growing quickly, on a per capita skier basis, it makes resort skiing look like an unruly rock concert, particularly when you consider that a significant percentage of resort skiers are beginners or infrequent skiers, who ski on the thinnest edge of what might be deemed "control." In backcountry skiing—particularly when it comes to travel in high avalanche probability conditions—the practitioners are by nature immersing themselves in terrain that is predisposed toward things about to go horribly wrong.

It's why the Utah Avalanche Center has its cheery slogan that is a knowing wink of the eye: "Keeping You On Top." Forecasters like Gordon seek to do just that, but even when their focus is confined to a designated mountain range—like Gordon's purview of the Uintas—it's impossible to know what's hidden on every peak, couloir, and runout to the canyons below. Gordon's charge in the Uintas covers nearly fourteen-hundred-square miles of the highest mountains in the contiguous United States, in winter a snow-blasted, meteorological battlefield. He can't post warnings on individual slopes, he can't dig snow pits beneath every threatening cornice, he can't draw a perfect line between affable twenty-eight-degree pitches and thirty-eight degree death traps.

Professional avalanche forecasters are, by necessity, an elite group whose experience in and knowledge of the mountains is exponentially complex. Their basic understanding of snow science is just the beginning, bachelor's degree level of qualifications. Ultimately, they aspire to master's and PhD levels which are earned over time as intimate knowledge of specific mountain ranges becomes necessary, involving the ability to quickly visualize isolated terrain features at individual elevations, and an understanding of how the day's wind and temperature are affecting snow on those slopes, and how this topography and weather data conspire across the compass rose.

All this incoming weather data, paired with that terrain familiarity, alchemizes into the advisories Gordon and his colleagues issue every day. But still, people die in avalanches. As much as Gordon and other veteran snow professionals try to remain objective in the face of deaths, their on-the-job experiences exact an unshakable emotional toll. In early February 2019, for example, Gordon investigated the avalanche death of a snowmobiler in an area called Chalk Creek. Over the years, Anita Gordon says, her husband had managed to compartmentalize emotions kindled by accident investigations. But not this one.

"It got to him that this guy will never go home to his wife and kids," she says. "The avalanche dragon doesn't know you have a home and job to go back to." The number of people killed by avalanches, relative to more run-of-the-mill lethal misfortunes—car crashes, pedestrian crosswalk rundowns, runaway home fires, gunpoint robberies gone bad, and a host of others—isn't high, but it's still devastating.

In more modern annals, deaths from skiing and deaths from avalanche are not usually reported together. An average of about forty people die every year in ski and snowboard resort accidents, according to the National Ski Areas Association. In the 2016–17 season, forty-four deaths occurred. With that season recording nearly fifty-five-million resort skier days, the fatality rate is less than one death per million lift ticket uses.

When asked about deaths from avalanche in the backcountry, most professionals quote an average of about thirty per season. But reports, and even some fairly rigorous studies of avalanche deaths, present an uneven picture. Montana State University houses an extensive collection of snow science archives. In October 2000, at Montana's Big Sky ski resort, a study of avalanche deaths was presented at the International Snow Science Workshop. Since 1950, the study reported, 593 people have been killed in the United States by avalanches, about twelve per year. But the study also said during that fifty-year period "the number of avalanche fatalities have increased significantly" due to more people in the backcountry.

Another study, presented at the 2010 snow science workshop in Squaw Valley, California, said avalanche deaths "soared during the 1990s," hitting thirty-six in the 2007–2008 season, "the highest number killed in the modern era (post 1950)." This study noted that 90 percent of victims are men. "For males, the twenty to twenty-nine age group suffered the most deaths," concluded the study. "For females, surprisingly, it is the forty to forty-nine age group." This suggests that young males may, into their late twenties, be deemed slow learners about avalanche dangers and are still testosterone heavy. Women, however, may reach a place in their mid-forties when they feel, because of extensive experience, overconfident about the odds being with them.

A number of US and European organizations aggregate numbers about avalanche injuries and deaths, but the exact number remains elusive. Several forecasters said there would have to be a person stationed at every trailhead to get a count on backcountry ski numbers, and there would have to be some sort of law requiring a report on every injury. Deaths are one thing, since law enforcement gets involved. But in a place that is a little bit lawless by both definition and desire, it's not easy to enforce reports of injuries. Moreover, many affected don't want to draw attention to themselves out of embarrassment or being perceived as endangering others.

THE AMERICAN AVALANCHE ASSOCIATION (A3), based in Bozeman, Montana, is the recognized certification board for the three levels of avalanche training many snow professionals go through. In addition to its quarterly magazine for members, *The Avalanche Review*, the A3 publishes—far less frequently—a compendium of case studies, titled *The Snowy Torrents*, in which avalanche accidents of the previous several years are analyzed. According to A3's website: "Growing demand increases existing pressure on the inherently dangerous avalanche industry."

In Europe there are several convening organizations around alpine avalanche culture, prevention, and response. One is the International Commission for Alpine Rescue (ICAR). Situated in Kloton, Switzerland, ICAR was founded in 1948 as an initiative of the Austrian Alpine Club. Numbers of deaths range across a broad stretch of countries, mountain ranges, and small Alpine communities with an extensive, historical mural of skiing triumphs and tragic deaths. True to their reputation of being precise, the Swiss produced an extensive report about avalanche deaths in that country, presented by the Institute for Snow and Avalanche Research in Davos. The report stated: "Of the 13,000 avalanches recorded . . . for the years 1936/37 to 2011/12, 3,540 avalanches involved 8,356 people. . . . During the last 76 years, 1,884 people died in 1,194 avalanches in Switzerland (annual mean: 25)." The Swiss numbers, while lower than the quoted average in the United States, reflect a number that floats above and below thirty per year.

Because the backcountry carries an air of mystery for many skiers and boarders, there are all manner of associated myths around it. Doug Chabot of Bozeman is particularly sensitive to things he hears that are contrary to the realities of travel in backcountry terrain. "You'll hear people say, 'Oh, I skied the slope but it didn't avalanche because I was light on my feet,'" he says. Or "'Ski it lightly,' you'll hear someone say, 'Ski it lightly, don't be too heavy on your skis.' And I'm thinking, 'My God, if that's the difference between dying or not, or getting caught in an avalanche, no one can thread that needle. . . . If it works for you, you just got lucky. Don't confuse luck with skill."

In the end, if the forecasters and avalanche centers don't get the word out to skiers, boarders, and snowmobilers in a timely and easy-to-understand format, all the data in the world doesn't matter. During winter, Gordon appears every Sunday at 7:35 a.m. on the Salt Lake City Fox television affiliate. He talks about the prior week's snowpack, about what's happening that day around the resorts and in the backcountry, and what's coming up along the Wasatch Range storm front. He believes putting a face, beyond the avalanche center's website, on the work forecasters do helps build their trustworthiness. Videos have become the virtual social-media campfire around which forecasters and skiers sit and converse about what's going on with the snow. Most are created by the avalanche centers, but some videos are sent in from skiers and boarders. Social media is the new dimension for communication.

In January 2012, while doing reconnaissance in the backcountry, Gordon and a partner triggered a huge avalanche. They decided to record it and put it on YouTube. The post went viral. "Instead of it being abstract, or a still photo," Gordon recalls, "what you got to sense and feel was the immediacy of it. I felt almost like a CNN reporter. . . . That was what was so revolutionary about it. It felt like the first time with a rotary phone: 'Holy Christ, there's a voice on the other end of this thing!'"

Chabot agrees that videos and social media have changed the game in tightening his center's link to their audience as well as to their ultimate mission. "It provides a megaphone effect. It goes out far and wide. We can just shotgun that out there. I know it saves lives."

8

SNOWFLAKES: WHERE AVALANCHES ARE BORN

As children, we are taught no two snowflakes are alike, an adage that's scientifically true. In kindergarten, we take paper, white as milk, fold it, and scissor the spine, cutting into it diamonds, squares, half rounds, and little pinnacles of star points. Opening the folded paper, we reveal the variegated snowflakes, our own creations distinct from our companions'. We are given snow globes to shake and peer into, containing worlds unto themselves that evoke the romance of winter's snowfall. We grow up perceiving snow as quiescent, an idyllic, white equalizer of the world, appearing like a wondrous gift one morning, hiding all the landscape's blemishes, as near as nature can deliver a peaceful, visual tableau.

But the formation and life of a snowflake is, in truth, an act of meteorological violence. It is, after all, a phenomenon of nature, and nature bends toward the arc of catastrophe. Hurricanes. Floods. Cyclones. Earthquakes. Lightning bolts. Hellfire. A snowflake can begin with only a microscopic mote of dust. A nearly invisible fleck of dried mud from the banks of Oregon's Rogue River. Sun-scorched dirt, like dark talcum, from California's Salinas Valley. The frenzied desert particles tossed skyward in a Nevada dust devil. Ash from forest fires and burning homes trapped in the downslope gusts of the Santa Anas, born in the Great Basin. Pollen from pine trees rimming the slopes of Lake Tahoe.

All this microscopic matter and more is swept into the atmosphere by icy winter winds bearing moisture, fine as mist. Swirling water, millions of minuscule particles, air about to flash freeze—a kind of fusion is about to begin. But even then, snowflakes are fastidious about what hosts they choose, and when they form. "Ice is very particular," observes NOAA's Randy Graham. "Only five percent to ten percent of particulates floating in the air can host ice. At minus seven degrees Celsius, ice will form on sea salt. However, a piece of vermiculite clay will only host ice at ten degrees or colder."

Water molecules seek out specks of flying dust so they have something to cling to. But a single pairing—one liquid molecule to a single dust mote—just begins the formation of a snowflake. Six to eight such duets must happen to form the core crystal. Then the hexagonal or octagonal flake truly begins to adorn itself, insatiably annexing more water droplets, reaching out its arms, the snowflake spinning like a pinwheel into a unique configuration of thousands of freezing water molecules. When first created, the flakes are pure, crystalline amalgams, each a tiny hall of mirrors, a prism as marvelous as if it were a living thing. Soon, a single snowflake has become its own infinitesimally small galaxy.

So complex is the conception of snowflakes, experts like Graham refer to tiers and structural characteristics usually governed by temperature. From 0 degrees to minus 3 degrees Celsius, snowflakes form as flat plates; from minus 3 degrees to minus 10 degrees Celsius, the flakes take on shapes that bear hollow columns, needles, and solid prisms. When the air is even colder (minus 10 to minus 20 degrees Celsius), plates form, dendrites that resemble proliferating tree branches, or, when bigger, ferns. Diamond dust crystals, the smallest snow crystals, are born in bitter cold, landing on one's shoulders like glitter in the sunlight, dispensed from the hand of God.

"Snow crystals," Graham says, "are like an atmospheric chemistry problem." Even if two snowflakes, under magnification, were to appear identical, they would still be different in atomic structure. Millions of them form together in the clouds, swarming, riding the wind, collecting more freezing droplets until they weigh enough to submit to the grasp of gravity and plummet toward earth. People talk about snowbanks, or snowfields, or slopes. But a foot or two of fresh snowflakes are really more like a colony. They reinforce each other, share a wintry ecosystem, like archipelagos gathered in a white sea.

In a mountain blizzard, the sort delivered to the peaks of Utah's Wasatch, wind is invariably present. It may help accelerate the flow of snowflakes to the ground. But it also can severely disfigure the flakes. With less wind, the flakes begin to connect to each other, linking their tiny arms as if in an icy white jigsaw. Such linkage and compaction can stabilize a snowpack, make it more resistant to a slide. But if caught in ground-level gusts, snowflakes begin to tumble. They furiously smash against each other, their elaborate crystal arms shearing off. Once elegant, they are now dull-edged, piled up against each other in drifts without architecture.

Environmental vagaries start to take effect. With temperatures spiking up and down, moisture may seep through the deformed flakes, creating a kind of snow cancer. The blunted snowflake stubs have lost their minute, perfect-locking features. Such snow has lost any affirmation of stability and is now more apt to break apart, like participants in a daisy chain suddenly ripping apart hands. All it takes now is surface weight—a skier, snowboarder, or snowmobiler—and the inherent mass of heavy new snow starts to break free, dragging with it tons of old snow from beneath. A rumble vexes the air. An avalanche dawns over the face of the mountain.

IF IT SEEMS THE SCIENCE of snow creation and behavior is thoroughly understood, the professionals say there is still a lot of mystery to it. "We're nowhere near learning all there is to know about snow," says Montana avalanche forecaster Doug Chabot. "I need to catch myself when I think that I'm figuring it out." When he gets out of the office and makes his forays into the backcountry, the first thing he seeks to learn is the temperament of the snow. "When you go into the field, you always have a question you're trying to answer. . . . Our job isn't to make educated guesses. Our job is to see what's really happening."

Chabot is a passionate advocate for digging snow pits in the backcountry. "If we go out there and dig a snow pit, and we deem that the danger's low," he says, "the next day the traffic will be fivefold, so we better be right."

When Utah Avalanche Center's Craig Gordon talks about snow, he refers to its life and its personality. "The snowpack to me has a life of its own. You watch it grow from prenatal to mature." A snowpack that is becoming more unstable strikes Gordon as an incipient delinquent. "It gets into its teenage years and you look at it a certain way and it gets pissed off." When he's skiing in the backcountry, Gordon says he can perceive the mood of the snow on any given day. He encounters deep snow that is hazardous, like surfing big waves that aren't clean and where shark fins pierce the water. Then a few days later, he'll ski a new snowfall that has a slightly "spongy, reboundable" quality to it as if with each powder turn the snow is levitating his skis a little higher.

"It's got life to it," Gordon muses. "When the snowpack has life to it, it becomes this third-dimension thing. It's yin and yang. It becomes a participant."

9

THE SNOW MONK OF LITTLE COTTONWOOD CANYON

As you thread your way up Little Cottonwood Canyon to Alta ski area, you pass multiple signs marking infamous mountain cliff bands and steep, craggy runouts. "Maybird" and "Tanners" and "White Pine Chutes," the signs tick off. As many as 165 targets within thirty-three critical slide paths over four miles have been mapped up the canyon. With the highest avalanche hazard rating of any significant route in the United States, the road is a skier's first black diamond run before he or she has even strapped on any skis.

So steep are the stone spires atop the canyon's walls, you are almost forced, the deeper you go, to fight off a faint bout of vertigo. Carved by an alpine glacier nearly twenty thousand years ago, the canyon emanates an air of historic nobility and contemporary menace. Blocks of granite, used by the Mormons to build their Salt Lake Temple, were quarried here. Legendary avalanches have swiftly shoveled surprised skiers into shallow graves. On signs that mark nearly every curve are the repeated words "Avalanche Area." In winter, the two lanes are a minefield waiting to detonate. As each new storm barrels in from the west, sucking moisture from the Great Salt Lake, snow arrives at the canyon's mouth like a cyclone. On such days, the slow, bumper-to-bumper motorists who strive to make it to Alta form a serpentine string of red taillights. It's known as the "red snake." As if one were captive in its belly.

One day, skiing Alta at mid-mountain, I watched a storm, black with snow, arrive far down the canyon and slam shut all daylight from the Salt Lake valley below. It swallowed the entire canyon in minutes. By the time my wife and I made it to the bottom, we were staggering victims in a frozen sandstorm. Witnessing all this from a perch at Alta is sixty-seven-year-old Titus Case. For forty-two years, he has patrolled the ski area's slopes, twenty-seven of those years serving as avalanche director. Though technically retired, Case still works avalanche abatement when needed. So intrinsic is he to Alta's operations, its history and ski culture, he is referred to by some as "the Snow Monk."

"There are a lot of people who come here to Alta and this is where they want to be the rest of their lives," says Case, who did just that when he moved to Alta after college in Vermont. "It's a unique place, not only because of the mountain and the snow we get. But the ski area itself is unique in the sense that it's not run by a big corporation. We don't sell real estate, or build condos. Alta just provides motor transportation to get up the hill on our lifts. . . . Day in and day out, there's no better place for skiing."

If Alta was the "laboratory" for the snow science work in the 1940s and 1950s done by Montgomery Atwater and Ed LaChappelle, it has also served as the testing ground for traditional avalanche mitigation techniques and the newest innovations. And Case has experienced them all. Sitting inside the patrol shack at the base of the mountain, he bears the trim, rangy build of an athlete at rest. He has the youthful, singular seriousness of a Boy Scout on a mission to keep people from getting hurt in the mountains. That's been at the heart of his labors at Alta. His voice can get almost somber when he talks about handling explosives, but then it shifts when he describes the use of bombs to blow up snow cornices at sunrise. There's a bit of a twinkle in his eye, a smile at the corner of his mouth revealing the vestiges of a boy's zealotry for cherry bombs and firecrackers.

"My first love was doing avalanche control work," Case says. "That was the direction I was always headed in." While he acknowledges that artificially triggered avalanches are the product of mitigation work, he sees avalanches as acts of nature in waiting. "We're just sort of helping along the process. We can't always wait for Mother Nature to have an avalanche naturally. Our job is to get the mountain open."

When Case talks about the work, he exposes the inherent romance of it. "It's attractive for a couple of reasons. One, you're up early, out on the hill. A lot of times you're riding the lift when it's still dark out." Alpenglow tints the peaks at Alta, Mount Baldy at just over eleven thousand feet, and the knife-ridged runs down Devil's Castle. Across Little Cottonwood Canyon, Mount Superior's own 11,040-foot face emerges from shadow. Case drinks in the chill morning air, about to unleash a sonic boom against the canyon walls.

"Watching big avalanches and being close to them when they break and go down the hill is an extraordinary experience." Like Craig Gordon, Shannon Finch, and other mountain professionals, Case speaks almost reverentially about snow and how avalanche mitigation has filled him with an awe that comes only after many years in nature's own winter monastery. "The other attraction is learning about the snow, snow structure. And why avalanches occur and what you're looking for when you're looking in the snow to determine when things are close to avalanching, or when they are stable."

At many resorts, avalanche mitigation is what's known as "snow safety." But Case and others at Alta feel this can be a confusing term, so years ago they switched to something less ambiguous. Now they call it the Avalanche Department. But Case makes it clear that these duties don't fall to just a small, designated group. At Alta, he says, if you are on the ski patrol, you are on avalanche assignment. "It's all hands on deck when it comes to avalanche control. It's full on. Everyone on the patrol gets involved."

CASE REMEMBERS HIS FIRST SEASON on patrol at Alta in the winter of 1975–76. Although he was eager to hit the slopes in his new patrol jacket, he and the patrollers didn't officially start until January 5—there was too little snow. But the next eight years revealed to Case the many skier's secrets of being a patroller at the mountain. Understanding those secrets included training at the US Forest Service's "blaster school" for handling avalanche explosives, a duty Case found especially engaging.

By then Case had become assistant patrol director, a role he held for three years. Meanwhile, legendary patroller Onno Wieringa had been running the avalanche mitigation program. A year after Wieringa was promoted to general manager of Alta, and Case assumed the directorship of the avalanche department—his ideal role on the mountain. "It's explosives and it makes a big noise. You got to have your thinking cap on the whole time. Handling explosives, going through the proper procedures and being careful.... It's serious work and you have to be all engaged."

Only a limited number of ski areas—among them Alta, adjacent Snowbird, Mammoth mountain in California, and Wyoming's Jackson Hole—have resorted to military howitzers for snow abatement. Virtually all other Utah ski areas perform avalanche work using thrown or planted hand charges, explosives hung from long poles, or the nitrogen gas-fueled gun called an

Avalauncher, which lobs a charge up onto steep faces or into chutes. Early Avalauncher designs were developed in the late 1950s at Alta by Montgomery Atwater, fashioned after baseball pitching machines. An Avalauncher's range and accuracy is perfect for early morning release of slides in resorts where the distance may only be four to seven football fields away; they can shoot farther, but because the shells are lightweight, stiff winds and longer distances can negatively impact the aim of pinpoint targeting.

At Alta, though, some of the most serious avalanche threats are more than a mile away, like the slopes of Mount Superior across the road that serves Alta. For that work, Case and his team roll out literal artillery—a 105 recoilless howitzer of World War II vintage. Such a gun can sling a nearly three-foot shell as far as seven miles, though at Alta the goal is more about precision placement of the shell. "In the middle of a snowstorm, you can blind fire a howitzer.... It's hard to beat artillery. It goes where you want it to. You can shoot it in a storm when you can't see your hand in front of your face," Case says.

To be authorized to fire such guns—weapons originally designed for another kind of war—Case and others become members of the Avalanche Artillery Users of North America Committee. Ski areas, Utah Department of Transportation gunners, US Forest Service employees, and the US Army all have members of the organization. If fighting avalanches with a cannon summons an analogy to "war," the avalanche controllers at Stevens Pass in Washington State took it one step further—they brought in an army tank for firing on dangerous snowpack on mountain slopes.

AFTER TALKING FOR A WHILE in the patrol shack, Case is eager to get out on the snow. Full sun is on the mountain, and the snow is softening, so we head up on the Collins Lift with a couple of ski patrollers. Case wants to show me some of Alta's newer snow mitigation technology. We stop in the shadow of Mount Baldy where an Avalauncher sits, aimed at the Baldy Chutes, an array of sometimes problematic, steep pitches known to funnel avalanches. Mounted on a steel platform, the Avalauncher looks like what it is, a small cannon of sorts, connected to a nitrogen tank that feeds the ammunition for propelling explosive charges up about fifteen hundred feet. A steel screen separates the launching barrel from the spot where a patroller fires the launcher. The screen was added years earlier, under a national ski area recommendation, after a charge exploded in a gun and seriously injured a controller.

Case points up to the left of a cliff edge leading into one of the chutes and draws my attention to a spherical gray ball that seems to be hovering like a satellite on the snow. It's called an O'Bellx exploder, a French-made device that can be remotely programmed by computer to combine hydrogen and oxygen in a blast that doesn't enter the snow like a bomb; instead, it delivers a concussive explosion in the air above the snow. The shockwave can be even more effective at releasing snow than a bomb, because heavy snow can insulate a blast as if it were inside a bomb squad explosion vessel. The O'Bellx was placed on Baldy at a particularly troublesome spot. One of its best features, says Case, is that its chemical tanks can be easily replaced by a helicopter lowering the O'Bellx to the foot of the mountain, filling its chambers with fresh chemicals, and efficiently ferrying it back to the mountain peak.

A similar, more infrastructure-heavy device is being used by the Utah Department of Transportation to mitigate avalanches above a main road at neighboring Snowbird and for avalanche mitigation above the valley's main ski area access road. It's called GazX and uses a large-mouthed pipe, curved to face the ground, and it emits a forceful belch of ignited oxygen and propane. Like the O'Bellx, it creates a shockwave in the air that loosens avalanche-prone snow and sends it careening down the mountain. The downside to GazX, Case says, is that it requires a tank farm and piping to hold the propane that feeds it, and the big emission pipes have to be mounted on permanent concrete pads to absorb repeated shocks when it's fired.

Case expects such new technology offerings to increase. "It's likely the Avalaunchers and military artillery will not be around forever." What will be around, he notes, are more skiers at Alta seeking to go into the backcountry. "One of the big changes is the equipment. It's changed the way people ski. When I was coming up, expert terrain was expert terrain. With the wider skis, it allows more people to go everywhere. People are going to get into every nook and cranny."

Dynamics other than wide skis have increased the patrol's range of oversight, Case says. High-speed chair lifts carry people up the mountain far faster, which means they can log more vertical feet in expansive terrain in a day. For some skiers, main runs become too crowded, enticing them to venture into side-country. All this, Case implies, isn't necessarily a good thing. Skiers fatigued by so many runs may place themselves in terrain they didn't foresee being so tough and above their skill level. Frequently, they don't bring needed knowledge or equipment.

Case and his fellow patrollers often encounter skiers who decide on a whim to "duck the boundaries because the backcountry is so attractive. When they leave a ski area like that they aren't properly prepared." In the patrol shack where Case had sat earlier there's a bumper sticker on the wall that reads: "Don't Believe Everything You Think." It speaks to another dimension of the ski world that Case and his Alta patrollers see and talk about—human judgment. That often applies in-bounds, but it is amplified by ten times in the side-country and backcountry.

"There are shortcuts and we humans use them to simplify things," he says. Where people feel beset by heightened threats—fragile snow cornices, steep pitches bearing cracks in the snow, wind-sculpted crust slabs—they tend to have less situational awareness because of psychological distractions like encroaching fear. We become, in Case's words, "blind to our blindness." Unexpected and deadly, the crack of an avalanche can suddenly fill the air. "Uncertainty is a large part of what we do. We deal with uncertainty. Yes, snow science is 'science.'... We know what makes an avalanche, but all that being said, we are constantly surprised when an avalanche occurs."

Many who follow Case's path after college to a ski patrol job last only a few seasons then gravitate to "real jobs." After he started at Alta, Case would visit his family in New York and his mother would always ask: "Are you still going to patrol?" His father, an avid skier who ran a lithography company in Rochester that printed advertising for large companies, would come out to ski Alta with him. "As it became obvious this is what I was going to do for life, I think he got a kick out of it."

Like any profession, Case concedes, there are sacrifices to be made. "There's a lot of hardships. When storms are in, you're up at 3:30 a.m. You have to make a lot of phone calls to patrollers. What's the forecast? How many shots [of the howitzer]? What's the afternoon like? If it's snowing, you have to get up the next day and do it again." He doesn't know how some patrollers balance the long hours of mountain work with raising families. "I don't have a family. I never got married," which may partly explain his nickname of "the Snow Monk."

Reflecting on his move from New England to Utah, when he was barely out of college, Case is more than at peace with his choice. "I was the kind of guy, when I left school, I had no idea what I wanted to do. My experience here at Alta is a dream come true."

10

SIGNING RISK CONTRACTS

In the mountains death by avalanche comes, as Everyman characterized the mindset before one perishes, when people have it "least in mind." Doug Coombs, one of the most experienced, formidable, and elegant extreme skiers in history, did not die in an avalanche while skiing the Polichinelle Couloir in La Grave, France. He wasn't even attempting a precipitous first descent of an elevator-shaft couloir. He was out with a group of friends, getting in turns on a sunny day. Trying to rescue a young partner who'd fallen over a cliff, Coombs slipped on ice and slid to his own death.

A friend, Jim Conway, who skied with Coombs at Bozeman's Montana State University and later served as a snow safety expert for the ski and snowboard filmmaker Teton Gravity Research, says he once warned Coombs about skiing too close to the edge. "I told him you can't keep getting on the edge. . . . You're getting too comfortable with the edge." Yet, Conway says, even now, this observation isn't "a criticism of Doug. Doug would have wanted us to learn from this."

You might think backcountry experts would be at the narrowest margin for an avalanche accident. Actually they constitute a large proportion of victims. Some have a high propensity for risk, which creates exposure. Some, skiing in unfamiliar terrain, overlook warning signals that escape the peripheral vision of their eyes and their minds. But most, skiing familiar slopes confidently, open themselves to avalanche dangers because they fall prey to evaluating risks, not based on the concrete clues before them but on scenarios their minds have drawn from past experience. If you've skied a particular slope many times and had a blast, dropping in today will undoubtedly lead to another blast. Unless, some find out, it doesn't. Such cognitive biases, based on stored memory, are called "heuristic traps."

ENGINEER AND RESEARCHER IAN MCCAMMON has studied avalanche deaths and the causes behind them. The loss of a friend, a highly capable skier, to an avalanche in 1995 caused him to delve into what might have played into his friend's death. It certainly wasn't a case of recklessness, McCammon knew. "Some claimed that Steve's death was the result of foolish risk, but I knew better," he wrote in a 2003 study published in *The Avalanche Review*. "Weeks earlier, I had shared a lift ride with Steve at Alta, and we had laughed about our climbing adventures years before. Things were different now, Steve said, and he told me about his wife and his beautiful daughter. He believed his days of being reckless were over, and the time for raising family had begun."

Experts in the backcountry, McCammon suspected, "have a large mental warehouse of experiences" that reassure them about their surroundings. Consequently, "they can unconsciously 'fill in the blanks' and construct a mental image of avalanche conditions even when their initial information is incomplete." To explore his theory, McCammon studied 715 avalanche accidents between 1972 and 2003. What he found was striking. "In cases where the trigger was known, 93 percent of the accidents in this study were started by the accident victims or by someone in their [group]." In other words, the victims weren't just the unwitting targets of natural disasters. They were participants in unleashing the fury. Defining seven avalanche hazard indicators—such as instability signs, obvious avalanche paths, and unstable, wind-loaded snow—McCammon determined that 73 percent of the accidents "occurred when there were three or more obvious indicators of the hazard." In fact, the data led him to conclude that "accidents where there was little or no evidence of the hazard prior to the avalanche appeared to be quite rare."

Another dynamic can reinforce problematic decision making in uncontrolled terrain: an almost incontrovertible cache of past evidence that things will be okay. "Most of the time," McCammon says, "the familiarity heuristic is reliable. But when the hazard increases and the setting remains the same, this rule of thumb can become a trap." McCammon explored other cognitive biases and environmental attributes. He wondered what effect commitment had on decisions: Does a high level of commitment to ski or snowboard a particular face lead to increased danger and accidents? Conversely, do ambiguous or even negative sentiments toward commitment open one to dwelling more on safety and decisions to shy away from skiing risky runs? Of 391 accidents McCammon reviewed, 253 cases showed victims had demonstrated a high level of commitment to achieving a stated goal and were influenced by things like approaching darkness, timing, or other compelling constraints. In 138 cases the groups had low commitment and the accident "typically occurred during the course of routine recreational activities."

Unless a person is skiing solo in the backcountry, a fundamental but pliable dynamic within commitment is that of one-on-one or group interaction. The effect this has on decisions derives from a complex array of perceptions stemming from perceived knowledge about fellow skiers' experience, personality traits, and even ego. If ski partners or members of a group are highly familiar with each others' backcountry skills and risk tolerance, then making real-time decisions—even those contrary to the prevailing grain—are easier. Less familiarity with a partner or group can raise communication barriers and dramatically mask commitment from person to person.

Avid skier Russell Costa, an associate professor of honors and neuroscience at Westminster College in Salt Lake City, has studied the psychology of personality interactions in the backcountry. Despite holding bachelor's and master's degrees, and a PhD in cognitive neuroscience, he says work in his psychology lab goes only so far compared to what he has witnessed in the backcountry when he straps climbing skins on his skis and begins trekking up a mountain face. "The most interesting things I've learned about human thought and human behavior are on the skin track. On a ridge, on the side of a mountain, in a snow cave."

Costa's studies, like McCammon's focus on risk and decision making but also on how leadership materializes in group dynamics and even what role gender diversity plays in more contemporary backcountry settings. "I think people are taking a little more risk," Costa says of his twenty years of experience in the Utah backcountry. "Or at least more people are taking more risk. Certainly there's always been those risk takers. But now there's a larger population of them."

This burgeoning pack of backcountry skiers and boarders, says Costa, has begun to respond to risk assessment through the filter of an added incentive—speed into the backcountry to gain first tracks. "There's now so many users. And there is sort of a pressure that if you want to ski particularly some of the iconic lines, [like] the south face of Mount Superior . . . you better be there on the skin track as the storm is ending. . . . I think there's a psychological pressure for people to get out there quicker and faster than there used to be." All this means a different equation for calculating risk. In the past, Costa says, people would give an unstable snowpack more time to "heal" and more safely adjust to a new load of snow. "But now the thought of encountering a tracked-up face creates more urgency to be the first."

New generational dynamics have come into play governing how groups interact in the backcountry. "There are online sites, I guess kind of like dating sites where people just sort of find in a Facebook group ski partners for the day. . . . Now people are more comfortable with meeting ski partners online. If they're traveling to the region, for example, or if they're new to town. And that's a very different group dynamic if there's a stranger there. Those are changing radically." The makeup of many backcountry groups is evolving. "When I started backcountry skiing," Costa recalls, "I would say the population was ninety percent men. . . . There's a lot higher percentage of females in these groups [now]. That's, of course, going to change how people interact." What isn't changing is the interplay of personalities when there are strong variations in peoples' risk aversion and risk tolerance. Throw communications styles and gender into that mix, and the result can range from sublime to disastrous.

As he forges ahead with his studies of backcountry behavior, Costa explores comparable psychological dynamics behind aviation accidents, in medical decisions, and even human motivations in gambling. But there are not always clear analogies. "People have lost money, then made money back and gone and gambled again. You can't do that with a risk that will kill you. Single trial learning doesn't exist in that domain. And I think that changes—for me anyway—it very much changes the psychology." The number and weight of psychological and objective variables makes it difficult to accurately gauge behavior. High commitment, whether through stubbornness or overconfidence, may overshadow or silence concerns about risk. Other backcountry human factors, like the time of day or an unwillingness to verge from a prearranged plan, can also stifle decision making.

In his 2003 study McCammon refers to the book *Snow Sense* by Jill Fredston and Doug Fesler, the Alaska Avalanche Forecast Center couple who mentored Bruce Tremper and others in avalanche education. In their work they explore the dangers of what they term "cow syndrome"—the rush "to get back to the barn" as dusk begins to descend in the backcountry. Another effect is "lion syndrome"—"the rush to be the first to get to a summit or a particular slope." In remarks delivered in 2009 to the International Snow Science Workshop in Davos, Switzerland, McCammon referenced a 2006 book by Sidney Dekker, *The Field Guide to Understanding Human Error*. A personality type is described in that work as the "bad apple," those people who, even in the face of clear hazard, make dangerous decisions and often draw others into that peril.

One safety expert at a major Utah resort addressed an avalanche study group in January 2017 about "terrain traps," gullies and treed runouts that funnel avalanches into narrow, lower mountain features where it's easy to become buried. Drop into a terrain trap, or follow someone into one, and you are ratcheting up the danger. "When you descend into a terrain trap and you've got ten or twenty skiers and snowboarders on the peaks above you," this expert explained, "any of them can start an avalanche. You have just signed a risk contract with all those people."

Is it possible to escape the temporal prison of heuristic traps—"bad apple" personalities, diminishing time, and onrushing darkness? Experts like McCammon, Tremper, Alta's Titus Case, and avalanche educator Craig Gordon believe it can be attained. But it requires a cognitive system or checklist. During his time as avalanche director at Alta, Case brought McCammon in to talk with ski patrollers about his studies. To this day, he refers to McCammon's acronym—FACETS—for gauging risks in the backcountry.

"F" is for "familiarity"—a trap that can move you to quickly conclude a slope is less risky because you've skied it many times before. "A" is for unquestioning "acceptance" of a decision to ski a pitch when there are obvious avalanche signs. "C" is for "commitment" and the need to carefully balance it with risk assessment. "E" is for the "expert," a halo that can blind a self-avowed expert to the simplest of threats. "T" is for "tracks" which can be interpreted as a determination to ski pristine powder no matter what, or even an erroneous assumption that skiing a slope with existing tracks suggests it was skied without incident and must be safe for you. "S" is for "scarcity"—scarcity of time, scarcity of untracked snow, whatever prompts you to make quick decisions without devoting yourself to the rigorous mental framework that could save your life.

All this is a systematic reflection of the bumper sticker in Case's Alta office: "Don't Believe Everything You Think." When shared by a group, this system removes the emotions and personality quirks that can silence some skiers, leading to "groupthink" that could produce a false sense of security and, unfortunately, severe consequences.

No avalanche accident embodied this concept more than the one at Tunnel Creek in Stevens Pass in the Cascade mountains of Washington State on February 19, 2012. Chillingly chronicled in a *New York Times* series, a group of fifteen highly experienced backcountry skiers and snowboarders, who had spent little time with each other, neglected to air their opinions about dangerous avalanche conditions. Because no one wanted to be seen as responsible for canceling elaborate ski plans, the group headed into the Cascades, dropped in on a forty-two-degree slope, and initiated an avalanche that slid 2,650 vertical feet. Three people died in the slide. Titus Case recalls that at least "two of the guys at Tunnel Creek" had worked at Alta's lodges; one of them died in the avalanche.

For his part, Tremper takes the mental discipline of a checklist and visualizes it in the backcountry. He uses the term common in behavioral science, "Ulysses contract," referring to the decisiveness that Ulysses exhibited when he ordered himself lashed to his ship's mast and ordered his crew to plug their ears so they could not hear the Sirens' song. Violating such a contract inevitably leads to disaster. Regarding risk contracts and checklists as they apply to heli-skiing, Tremper says, the best guides have a threefold system for planning the day's skiing. Once they've compiled snowfall data, wind and temperature forecasts, and their insights into mountain geography, they map ski terrain into three areas—green for definite skiing, yellow for possible, and red for terrain "not to be guided under any circumstances." And then they stick to it.

Tremper says skiing fresh powder in the backcountry and evaluating terrain that's steep and avalanche prone versus terrain that's lower angle and far safer is a bit like trading stocks. "You're balancing greed and fear all the time," but adhering to a checklist makes finding the balance easier and shrinks the risk. "The system protects you from the errors that all of us are going to make. . . . We have these optimistic biases. We think we are better than we are."

11

THE DRAGON'S LAIR: BURIED ALIVE

On April 1, 2011, the world-renowned mountain photographer and rock climber Jimmy Chin was buried alive in a Tetons avalanche. Chin survived this near-death experience and later described it in his private journal:

> *I have one last glimpse of where I am going as I get drawn into the darkness. Hope fades and fear rises. . . . Then comes the weight. It pushes down. It compresses. It is more and more and more and more. . . . It is unbearable. I hear myself roar from a place I knew a long time ago. It is primal. It comes from my stomach and into my chest. I hold onto my body. Bracing, bracing, tightening for impact. The impact never comes, but the weight gives me no release and I feel my chest compressed and crushed. No chance to breathe. No chance to expand my lungs. It is dark and it is dark.*

CHIN ALLOWED HIS NEAR-DEATH ACCOUNT of being caught in the avalanche to be shared on the internet in July 2013, "to inform but also keep the memory fresh." Years later, he would go on to produce, with his wife Elizabeth Chai Vasarhelyi, the Academy Award–winning film *Free Solo* about free climber Alex Honnold and his mind-bending ascent of Yosemite's El Capitan.

Chin recounts being submerged in deep snow during the initial slide down the mountain. When the avalanche ultimately came to a stop, he was upright with snow barely two inches from his face. He could breathe, and he started digging himself out "alive and somehow, uninjured." The survival and death rates for people caught in avalanches are remarkably predictable. The moment Chin felt his skis going into a freefall down the mountain, he had about a 50 percent chance of the outcome he experienced. Partial burial, the ability to gasp air, and even use his hands and arms to begin digging himself out were what saved Chin. It's likely that Chin, being so experienced in the mountains, did some instinctive things that helped him survive, things he may not even have realized—swimming his arms in the snow, fighting for the surface and sunlight, mentally refusing to submit to death. But he was also lucky.

As recounted in his journal, there was another moment in which he feared a different outcome: "I am slowing down, but the weight is still too much to move. I am encased in concrete. They will never find me."

Imagine an avalanche hitting an area where eight people are skiing. Half of them may become partially buried and escape danger entirely, but for the unlucky four caught in the avalanche, odds are pretty explicit. Roughly, one of them will be smashed into rocks and trees or swept over a cliff edge, dying quickly from blunt-force trauma. Such injuries include severe internal bleeding from the liver, spleen, and other organs as well as fractures or total severing of legs, arms, spinal cord, and skull. Another three victims will slowly suffocate from total burial in the avalanche. They may not be carrying an avalanche beacon, or may be alone. They may be too far away from any potential rescuer and have companions who are also buried.

A process will begin for those buried that, like the odds of living or dying, is equally foreseeable. Here, the hourglass—really, about a thirty-minute glass—begins to steadily feed sand into its lower chamber, its timeline divided into three distinct periods: the first is about ten to fifteen minutes, the window many die in; the second fifteen to thirty minutes might still offer life depending on snow compaction and air; the final period is a rapidly dwindling, tiny reservoir of hope, so few grains they can each be counted as they empty to the end of the chamber.

WHAT GOVERNS A VICTIM'S SHORT lifespan under avalanche snow is a combination of panicked human reactions mated with physiological mechanisms that no human, no matter how experienced, can overcome. Chin mentions the feeling of being "encased in concrete." It's the most common description of being straitjacketed in snow. Avalanche snow, once it gathers momentum, is usually moving like a freight train down the mountain, accelerating to fifty or even eighty miles an hour. Craig Gordon of the Utah Avalanche Center described the Uintas accident in which backcountry boarder Jeremy Jones's legs were broken this way: "It's sort of like being strapped to the front of a locomotive, and of course you're in the receiving end of everything that the snow is hitting."

As this massive amount of fast-moving snow consumes energy and tightly grinds together snow crystals, they become smaller, more fluid than solid. Heat builds up. A rising temperature causes some snow to marginally melt and it compacts more, and that effect, combined with millions of bonding crystals, leads to slabs becoming heavier and heavier. Under those slabs and surging snow, a body begins to sink deeper. Above, trees are being snapped, loose rocks are gleaned from the slopes, adding weight to the mass. When the slide finally stops, any heat trapped inside the avalanche suddenly starts to dissipate and, as freezing air merges with the moisture and tiny crystals, the snow refreezes. Its consistency is like a slurry that sets up to harden in seconds, as surely as concrete.

Unable to move, you panic. It's dark beneath the avalanche. Almost certainly, your mouth and nostrils are filled with snow. Eyelids and ears ripped at by ice shards may begin to swell and cause pain. At first, there is the instinctive urge to move your extremities, to fight. But once you realize movement is impossible, the fight is just a waste of whatever strength you have, so a kind of submission sets in. Uncertainty about perceived injuries to other areas of the body stimulates a new wave of panic. The inability to see or hear whether a companion or onlooker is coming to help creates a deep sense of helplessness and dread. Desperation overwhelms any judgment about survival. By now, you have used up three to five minutes. Robbed of the ability to fight for self-preservation, your focus turns to breath. It's the primary remaining link to life. Breath becomes everything. Only soon, it will become death's stalker.

Suffocation under an avalanche is different than drowning—and not only because dying underwater usually happens in two minutes or less. Under snow, chemical imbalances begin to take place in breath and the bloodstream, a process in which oxygen is leached away from the blood—asphyxiation. As if that's not enough of a challenge, in the sepulchral-like grasp of snow, your body temperature quickly falls, leading to hypothermia. Like an automatic physiological switch is flipped, your body's reaction to the extreme cold is to summon blood from the extremities to the thoracic cavity to protect the heart and other critical organs. The body's programmed impulse is a logical one: blood brings oxygen to the chest's organs and, just as essentially, to the brain. Your body is fighting for survival even if you can't mount a struggle.

Meanwhile, as you think about breath and try to manage it, the hourglass is now at about five to seven minutes. Suffocation in snow isn't caused by lack of air. Yes, the compacted snow may have left only a very small cavity around your head, but there is air within the snow. It's the overcompensating efforts to extract that air that may accelerate death by asphyxiation.

Respiration becomes your adversary. As you exhale warm breath, it interacts with the chilled layer of snow inches from your mouth and face. Snow crystals begin to form a thin mask of rime around your face—a mask that becomes a membrane blocking the easy flow of air. When you breathe in the diminishing air within this icy mask, you exhale carbon dioxide. A buildup of CO_2 starts. With each intake of air, more CO_2 is being ingested into the blood. As this concentration of CO_2-heavy blood is absorbed by the heart and other organs, you become confused. Another anatomical switch has been flipped. Because of the threat the brain perceives in this tainted blood, your very impulse to breathe is reduced. You start to feel drowsy. Some people who have survived an avalanche burial for this long describe a peacefulness, an acceptance, even articulated thoughts of good-bye wishes for loved ones and regret at the prospect of not seeing them again.

The hourglass is now at ten minutes or more and the preponderance of victims are at the edge of death. True asphyxiation has descended. The heart rate slows. Blood pressure drops. Cold infuses the cell structure of the skin and capillaries, and those deadened body protections now forfeit any insulation of the core organs. If you survive to fifteen minutes or more, a combination of advanced asphyxiation and creeping hypothermia will combine to end your life at the outside limit of twenty to thirty minutes. But most people die well before they reach that thin margin.

JIMMY CHIN LIVED TO WRITE about it. Just as victims on the cusp of death enter a mental realm that's philosophical about life's truest riches, so survivors, like Chin, are moved to weigh life's most essential graces. "There is a strange clarity for me," Chin wrote in the immediate aftermath of the avalanche. "Life's priorities are stacked perfectly in front of me. I know it is clarity I need to remember, that I can never forget."

Meditations on freezing to death, whether from an avalanche or in the Antarctic cold, vary in their evocations of bringing life into a final crystalline focus. No diarist, however, has captured this thought better than Apsley Cherry-

Garrard, author of the adventure masterpiece *The Worst Journey in the World.* "If the worst, or best, happens," Cherry-Garrard wrote, "and Death comes for you in the snow, he comes disguised as sleep, and you greet him rather as a welcome friend than a gruesome foe."

12

IN THE SELKIRKS: BIG RED CATS

As nurse anesthetist Ben Rabe prepared for his ski trip to Canada's Selkirk Range in southeastern British Columbia, he made a purchase epitomizing his excitement, a new pair of Armada Magic J skis, and one that stood as a metaphor for his anxiety, carbon fiber–shafted ski poles with breakaway wrist straps for avalanches. It was an uneasy equation that probably could have been applied to any of the eleven other men preparing to accompany Rabe on the five-day trip in January 2017. After all, like Rabe, most of the group dealt daily with a certain level of professional anxiety—his brother Jake was an emergency room doctor and four were dentists, like Rich Nydegger. The walks of life from which the men came said much about their sense of purpose in the world—professionals, family men, tightly bound to their communities and close circle of friends.

Once on the trip, Mike Crowton remembers having an "unsettling feeling" but dismissed it, "chalking it up to old age and three previous days of good, hard skiing." Crowton would later learn that his son, who was in Australia, thought about his father on the ski trip and "prayed fervently for my safety." All twelve were highly experienced skiers and snowboarders—indeed, handpicked for their expertise in the backcountry. None of the group wanted novices along who might impede their forays into the backcountry. Or worse, a rider who might get injured and require evacuation and the abrupt end of a day of snow-cat skiing.

Another trait the men shared: all were devoutly religious. If anything might have dispelled their apprehensions about heading into the backcountry, it was their faith that God was watching over them. Creighton Green had a deep belief in both God's protection and that of his own guardian angels whom he visualized as his deceased older brother and grandparents. These men were proud of their faith and not shy about publicly voicing it, even moments before dropping into a steep ski slope.

But on January 19, 2017, in the midst of an afternoon of skiing with Big Red Cats—a backcountry operation near Rossland, British Columbia—the men found themselves in a ten-hour battle against avalanches, vexing gravity, snow the consistency of quicksand, freezing night temperatures, and a treacherous mountain rescue that summoned every bit of their religious resolve.

THE SKI TRIP BEGAN MORE like an expedition than a five-day vacation to Canada. Ski carriers, boot bags, duffels, and suitcases were heaped together. Nine of the twelve men flew on January 15 from Salt Lake City to Spokane, Washington, where they met up with Jake Rabe, whose brother Ben was also on the trip. Two of the men had flown ahead to help with travel arrangements. From Spokane, the group rented two large vans and drove into British Columbia, planning to ski with two different snowcat operators.

Nydegger, a snowboarder and father of five from Sandy, Utah, recalls that the group initially needed to bond. While some members were related, or had known each other since childhood, others were meeting for the first time. Ben Rabe shared his "stoke" about the upcoming trip with his son's soccer coach only to discover the coach was on the same trip. As Nydegger remembers, after the first two days, it was as if all the men were instant family or long-time friends. They would be working with Valhalla Powdercats, located, as its promotional messaging says, "deep in the heart of the legendary Selkirk Mountains." The terrain, Valhalla's website boasts, offers "seasoned skiers a hearty selection of high alpine bowls, steep, old-growth forests, and endless, untracked Kootenay powder all day long."

With a shortage of new snow, the men still found Valhalla's skiing good, "even great," as Nydegger characterized it. To him, it had a variable quality, some of it loose and granular, snow you sunk into at every turn. "It was weird," Nydegger remembers. "It was really sugary snow. You'd go off a run or jump and land in snow up to your chest. It was not supporting snow." The two days had given the men a chance to get to know each other's personalities. By the end of the second day, they were so emotionally high from fellowship in the mountains, Nydegger recalls everyone "singing at the top of our lungs" as they headed back to a lodge.

One of the Valhalla snowcat guides called "Nez" was driving the large van, and he began honking the horn to no purpose as if providing a beat to the men's singing. In the van's wake was an image, painted on the vehicle's rear, of a huge skull and crossbones. Jake Rabe had put together what he called a "Nez playlist." Nydegger says there was Snoop Dog, Earth Wind & Fire, and Biz Markie's "Just a Friend," the latter song adopted as the trip's anthem. These early days of the trip had allowed the men to revert to their teenage selves, a choir of innocents with only snow to worry about.

DAYS BEFORE THE TRIP, POWERFUL snowstorms had hit the California Sierra and Utah's Wasatch, but they had stayed south of a high-pressure system blocking storms along the Canadian border. In the Gulf of Alaska, systems continued to churn into the atmosphere. Dense with moisture, one of the systems skated past the low-pressure ridge and slipped into Canada. As it hit the frigid air of the Selkirks, a heavy snow began to fall.

The next day, January 18, the men were scheduled to ski and board at Whitewater, in Nelson, BC. When they arrived that Wednesday, it was still snowing. Nydegger estimates that between the night's snowfall and what continued to descend, a foot to as much as two feet accumulated. At first ambivalent about traveling all the way to British Columbia to ski a resort, the men found the conditions so perfect they reveled in the deep powder. But as the day wore on, Nydegger says, "being in a ski resort, it obviously gets tracked up quickly." Heading back to lodging in Nelson, they anticipated the next day's snowcat skiing where thousands of acres of mountain pitches would be covered in new snow.

"I think we were all really looking forward to Thursday," Nydegger says, "with Big Red Cats with this new snow. I know I was."

FROM THE BEGINNING OF THE trip, the men had been practicing backcountry protocols for being caught in an avalanche. They had checked their beacons, performed practice probe pole strikes, shoveled as two-man teams to uncover imagined bodies. Just as singing and teasing each other had stimulated a bond among them, the serious simulations of avalanche rescue had created ties that might fortify their fortunes in the backcountry.

So early on Thursday morning, it was no surprise to them that they were issued Mammut "Pulse" avalanche transceivers when they arrived at Big Red

Cats loading spot. The group began avalanche response practice. Skier Clint Iverson recalls the repeated drills the men went through. "During the training, we had searched for buried beacons, used a probe to find what was buried and practiced proper techniques on how to uncover and dig out buried objects with our shovels. We felt pretty prepared."

Enjoying its eleventh season the day the group arrived, Big Red Cats is located about a two-and-a-half-hour drive from Spokane. The snowcat operation, a twenty-minute drive from Canada's Red Mountain ski resort, bills itself as "one of the biggest cat operations in the world" with the "largest snowcat road network in the world." Even compared to the largest ski resorts and many backcountry reaches, the Big Red Cats terrain is vast. It spans 19,300 acres divided by eight mountains. The operation claims to have more than five hundred named runs as well as "a bunch of unnamed runs." Its website says that when ski guides are hired there, it takes them at least two seasons of skiing every day to become fully knowledgeable about all the acreage, and even then guides have not skied certain runs.

The operation is owned by two Australian transplants to Rossland, Kieren and Paula Gaul. Kieren is lead guide, while Paula is president of the company, office manager, and "the real boss," as the website notes. The Gauls opened Big Red Cats in the winter of 2006–2007, with about seventy runs on a single mountain, Mount Neptune, and two snowcats. Over time, they expanded to more mountains and terrain, including adding, in their third season, Mount Mackie, a 7,070-foot peak, which became known to the Gauls as an avalanche-prone area when snow conditions were variable.

According to the website history of their operation, the Gauls recall the first season with Mount Mackie as "our worst avalanche year." Snowpack was so unpredictable, veterans of the area deemed it "the most unstable that it had been in about thirty years." They reported: "In some ways this was our toughest year. We had our most unstable snowpack—and our first and only full avalanche burial. This was a big wake-up call. Our guest was physically unhurt—but really shaken, as we all were." The Gauls, whom customers have praised for the priority they place on safety, acknowledged that the business of snowcat skiing does not operate without risks. "We can't eliminate the risk in what we do," they wrote, "but we made some mistakes on this day that we try to keep learning from!"

As that season wore on, the Gauls continued to explore their newest mountain, posting a report that "the west face of Mount Mackie was amazing. . . . We skied the avalanche paths on the east face of Mount Mackie, and we discovered the 'Shorties' area—short but incredibly steep runs which are always fun." In the ensuing seven seasons, the Gauls' business expanded significantly. Snowcats, small buses to transport skiers from the check-in office at Red Mountain, and a small restaurant named Fresh were added. Even music and insulation were introduced to the snowcat cabins.

Then the season of 2016–17 arrived. South of the Canadian border, fierce storms were hitting the Sierra and the Rockies. In British Columbia it was a winter of shifting conditions. That January in particular was an erratic month for the snowpack in the southern Selkirks. In the first seventeen days of the month, Red Mountain, just a short drive from the ski terrain of Big Red Cats, received only three days of snowfall, according to archived snow records on the website OnTheSnow.com. On New Years day, Red Mountain resort saw five inches of snow accumulation; the next day, another inch fell. Then a drought of six dry days occurred before another five inches fell on January 9. After that, another stretch of dry days ensued through January 17. From January 1 through January 18, there was a total of eleven inches of snow.

BY THEN, NYDEGGER AND THE group had skied two days with Valhalla Snowcats. On that Wednesday, they awoke to fresh snow and headed for Whitewater ski area; five inches of new snow was reported for that same day at Red Mountain. Nydegger and his companions reported seeing at Whitewater much higher accumulations than the five inches at Red Mountain, fifty-eight miles distant. Because of the way storms congregate differently around mountain peaks and valleys, it's possible for snow totals to vary widely in the same region. Eighteen inches of fresh snow was reported for Whitewater that Wednesday by OnTheSnow.com. The next day, January 19, Whitewater received another twenty-two inches—almost two feet.

Although Whitewater had seen more snow than Red Mountain the first half of January—it recorded forty-four inches total through January 16—it, too, had seen a drought of five days just before the men arrived. What's evident is that the "unsupporting" snow Nydegger remembers snowboarding at Whitewater was the deep new powder lying atop the underpinnings of January's early snowfall, old snow that had been decomposing before and during the five days without any accumulation. That Wednesday, and continuing into the night and early the next morning, snow fell over the Selkirks and Rossland, British Columbia.

By the time Nydegger and his crew arrived at Big Red Cats, a foot to possibly two feet of fresh white powder crowned the mountains.

FOR NYDEGGER AND HIS ELEVEN friends, that Thursday morning dawned as a dream. "Deep snow in beautiful, gladed trees," he recalls. At the start, they had been introduced to two snowcat guides who would be with them for the day. Lead guide Kauri Howell would typically "ski cut" or test the snow with a first run. He had been with Big Red Cats from the start. Not only did he guide, but in the summers he worked to open up more terrain by identifying tree cutting for added cat roads and downhill runs. "Kauri knows the Big Red Cats area as well as anyone and better than most," the operation's website notes. "Kauri is focused on safety and fun with guests."

The other guide, Ari Gore, known to them as "Gore," was to serve as their tail guide, skiing the final run after the twelve men had descended in pairs of two. Buoyed by the sight of morning's fresh snow, and the anticipatory mood over soon being ferried into the untracked backcountry, the men executed their avalanche beacon practice and other preparations. Everything went smoothly.

Getting to the first run of the day typically requires about a forty-minute ride in the cat. The first runs were down Mount Neptune, runs that participant Jared Flitton recalls as "steep, deep, and long." But as the morning went on, a decision was made to take the group up into the Mount Mackie terrain—a longer ride but an area Kauri apparently felt matched the group's skill level. Validating Kauri's judgment, all the men had been skiing and boarding aggressively. No one had backed off from steep, treed runs. Mount Mackie was known for some of its tight trees and patches of cliffs. It was expert level, what Big Red Cats sometimes called "über level" slopes. With the group's extensive backcountry experience, no one was inclined to veer from the plan: they would get in a few mid- and late-afternoon runs on Mount Mackie and then head for the Big Red Cats shed.

"This was the first day this peak had been opened all year," recalls Clint Iverson. "I felt like we had pushed our guides to take us to the most technical slopes they could find for our group, which consisted of very capable skiers." It had become apparent to the skiers that the guides, in particular Kauri, were sensitive to the day's snow conditions. "We were all instructed at the beginning of the day that the snow was not as stable as they would like," recalls

Tyson Cox, "and we would have to be very cautious." Skier Chris Bond echoes this memory: "Our guides seemed a little concerned about the snow consistency and were constantly checking the snowpack."

Each skiing with a partner, the men made tracks on Mount Mackie. Jared Flitton, who had skied the mountain a year earlier with Big Red Cats, recalls that one of his previous year's runs had left him "relieved to make it down" after he'd seen a cornice break from the peak and set off a big avalanche. Now, a year later, he was sure his lines down Mount Mackie showed a renewed confidence. "Truth be told, I was looking forward to a little redemption."

The afternoon plan hit a setback. So challenging was the skiing, and so large was the group, that two men became separated from the others. Chris Bond and Corey Penrod had disappeared into the trees, while others waited by the snowcat at the bottom. When more time passed and the two did not appear, the group went into search mode. The lead guide, Kauri, instructed Gore to begin skinning up the mountain to see if he could spot the two missing skiers. Kauri joined eight of the men to head back up the mountain in the cat so they could ski down and search for the missing members.

Disembarking from the snowcat and huddled at the top of the mountain, Brandon Flitton asked the assembled group to engage in a prayer for their lost friends. "We made a small circle," Clint Iverson remembers, "and asked our Heavenly Father to help us find our lost skiers. . . . I specifically remember feeling the spirit strongly during the prayer." Off to the side stood Kauri. Hearing the men pray, he decided to join in. "It was surprising but really great to see our guide, Kauri, ski over to join us in prayer," says Iverson. Below, Ben Rabe and Jason Bond waited and prayed, just as the men on top were doing. "It felt like an eternity to Jason and I," recalls Rabe, "who were taking turns yelling and blowing our whistles."

The nine men who had bonded in prayer prepared to start down. "At this point," says Creighton Green, "I started to realize just how awesome our group was. I remember my eyes filling with tears as I felt the spirit confirm to me that we would find them." With avalanche beacons in hand, Kauri and his eight clients fanned out.

And then, suddenly, the two missing men were found. Realizing they had skied too far to the left, they had stopped on the mountain, hoping they would catch a glimpse of other skiers and hear them. At the bottom, everyone convened in the snowcat to eat a quick lunch while the vehicle clawed its way back up the mountain. "A sense of peace and joy continued with us as we began our travel to our next adventure as a complete group," says Brandon Flitton.

Reunited, the twelve men prepared for their final, planned run. It was about 2:30 p.m. They had already made knee-deep turns through the wealth of flying powder; they had even posed for a group photo in front of the Big Red Cats snowcat, as if in celebration of their answered prayers. Creighton Green remembers: "I had mentioned to Kauri that the snow in front of us looked killer. It was untouched and deep. He said, 'Interesting choice of words,' and began telling us how they had a snowmobiler in that area that died in an avalanche. They made it to him a little too late, and I could tell Kauri didn't want to talk about the details."

Several of the skiers admit they were feeling uneasy about the final run, but no one spoke up. "We're a bunch of alpha males," says Nydegger.

Pointing his skis down the mountain, Kauri dropped in. The men watched him link tight turns down into a stand of trees. When the lead guide stopped, he radioed up to the sweep guide, Gore, and told him to send the men down in pairs, as Jared Flitton recalls, "spaced out about twenty seconds." Green and Cox were the first partners to go. Green recollects "some amazing turns in deep powder and evenly spaced trees." When they spotted Kauri, the pair drew up to him within a few feet. Standing there, Green and Cox could see a pitch a little farther down that opened into a large bowl, where according to Green, "it was smooth and mellow without any trees for about one hundred yards."

But Green also had noticed a patch of exposed rocks to the left. As skiers came into sight, he and a few others called out warnings. As they dodged the rocks, a few skiers stopped below where Green and Kauri waited.

Jake Rabe, the Spokane ER doctor, was one of them. He watched as Corey Penrod made his way down the first pitch, gathering speed. "I skied down and went off a little rock that kind of threw me off-balance," Penrod recalls. "I caught my balance and turned sharp left and barely missed running into Jake, who was standing still looking up the mountain. I remember after I stopped, I fell over and started to laugh." Penrod fixed his gaze on Jake's face. "I looked up at Jake and said, 'Sorry bro, I was coming in hot.'"

Penrod didn't see any humor in Jake Rabe's face. "Jake's eyes got big as I was saying this. His eyes were looking behind me up the mountain."

FOUR SKIERS WERE DESCENDING AS they heard an audible *woompf* and a large crack split the snow about twenty yards above. Still making turns toward the group below, Jason Bond screamed: "*Avalanche! Avalanche!*"

Snow, in huge blocks, rolled toward them, coming fast and violently. After a day of what seemed heroic, even spiritually elating skiing, the men and their guides now witnessed a mountain that had ceased its welcome. It had flown into a murderous rage—and it was coming for them. Iverson recalls: "The very thing we came to Canada for—fresh, soft snow—instantly became the enemy." And Green remembers: "My heart was racing and beating out of my chest. The magnitude of the avalanche was unlike anything I'd ever seen or felt."

Cox, who had been one of the first skiers down, was in a spot where he was "lucky and able to stay on my feet." But he glanced below to check on his fellow skiers. What he saw sent adrenaline through his body, every second growing more horrific. "The first person I saw start really sliding was Jake [Rabe]. He was instantly swept off his feet and going headfirst down the mountain in car-sized chunks of snow. I then saw Jared [Flitton] trying to ski out of it, but he got caught up and sort of dove head first sliding down on his back."

Within five to seven seconds of the start of the avalanche, Green saw "these brothers of mine" being carried toward the trees beneath the open bowl. "They had picked up a lot of speed. Upon impact to the trees, all of the heavy snow on the countless tree branches fell to the ground, while the sliding snow burst through the large and small trees creating an insanely large, white cloud. Almost like an explosion. At that moment, they were **GONE**." Nydegger remembers "a couple of guys tomahawking down, and then they were gone. It was the most sickening thing." Jared Flitton fought to stay above the rapids of surging snow, which seemed to drag him back under each time he gulped air. He saw light then was submerged in darkness. "I was moving faster and faster as the snow carried me down the mountain like a raging river. I could see nothing but white. . . . One second I would be choking on snow as it rushed into my mouth, the next I was yelling out of pain as I slammed into tree branches and rocks. . . . In those few seconds I thought that I would hit a tree or a rock and that would be the end. There was no way to control where I was going, no way to see what was ahead. Just the fear that at any moment I would be gone forever. And then I stopped."

CROUCHED BENEATH THE ROCK OVERHANG that earlier some skiers had been yelling warnings about, Mike Crowton heard "the sound of rushing snow over my head! . . . As the sound quieted, the last of the slide grabbed my skis and started pulling me downhill." Dragged into part of a clearing, Crowton stared across the small valley and felt, not the deceleration of the slide but a second avalanche suddenly break. He spotted the lead guide, Kauri, being shuttled down the slope. "He swims like his life depends on it," Crowton recalls, "and stays on top. I lose sight of everyone at this point."

Jake Rabe, who had been one of the first men sucked into the huge wave of snow, recalls almost nothing about the slide. Slammed into a tree that split his helmet in two, he was totally buried. In the aftermath he has tried to reconstruct what happened. One thing he is sure of is the seismic ferocity of the avalanche. "Based on the distance traveled and average speed of over eighty miles per hour, I [was] carried down the mountain in approximately ten seconds.

Seven of the twelve men had been swept below in the avalanche. Nydegger, who was one of the five who largely escaped the slide, was one of the first to spot Kauri after he had been seen swimming in the snow like a man trying to survive a rip current in the ocean. Now Nydegger could see the guide had lost his skis. "I remember him yelling, 'Get me my skis!'"

All across the mountain, panic set in. Fourteen men were strewn up and down Mount Mackie. At least seven were totally or partially buried, two of them three hundred yards down the mountain in a fight to breathe. Of those who weren't under the snow, most were injured, some seriously. But at this point, all Nydegger and the others in the group who had evaded the worst of the slide understood was that they had to begin searching. As experienced back-country skiers, they were aware of the stopwatch that had been set in motion for death by asphyxiation.

Nydegger recalls "pulling out my beacon and switch[ing] it to 'search' and that was sickening because the beacon was going berserk." The problem he and others surmised was that everyone not under the snow had to switch their beacons to search mode, otherwise any beacon still on "receive" would appear to be a buried body. With such a large, spread-out group, attaining that swiftly was precarious at best.

Well below Nydegger and the others, Jared Flitton "was stopped but I was buried. The relief of stopping was quickly replaced with panic from being under the snow and unable to breathe. My mouth and nose were full of snow. I could feel it choking me." He tried to move, but his entire body was locked inside a casket of snow. Then, from the darkness over his face, he felt a tingle. One of his hands was barely sticking through the crust. "I began to scoop the snow away from my head and face with my hand, only being able to move my wrist back and forth. I was able to uncover my face down to my chin. Just as I was able to see out from the snow, I threw up the snow from my throat and mouth."

Breathing in short gasps, Jared Flitton was imprisoned in the heavy snow blocks. Only able to peer skyward, he saw brilliant sunlight and tall pines, their branches laden with snow. For a moment a happiness filled him. "The incredible feeling of happiness to be alive was fleeting, as I quickly began to think of everyone else. I could not hear anyone. . . . I screamed over and over for help, but the voice that I so desperately wanted to hear did not follow." His thoughts, uplifted by the light bathing his face, raced ahead into a place of darkness. *Where is everybody? Everybody else must be buried. Everybody else is dead. I'm never going to get out of here. I'll never see my family again. I survived that horrible avalanche only to die from hypothermia.*

"I began to cry. I yelled for help over and over, and I pleaded to God to somehow save me."

UP THE MOUNTAIN, AN EERIE silence hung over the stunned men who had escaped the avalanche. If there were people still alive below them, they could hear nothing. Then a sound grew, emanating from under the snow—the insistent, percussive beat of avalanche beacons, beeping in search mode, frantically hunting the air for electronic links to beacons buried beneath tons of debris.

Clicked back into his ski bindings and with his own beacon in one hand, Kauri summoned Cox and Nydegger to help search. Soon after starting, they spotted Brandon Flitton off to one side. "Kauri and I continued down through the trees," Nydegger says. "We were getting signals bouncing all over the place. We also kept seeing poles and skis randomly down the hill. We came into another clearing and I saw Ben [Rabe] pop out of the snow."

Then Corey Penrod emerged. Cox felt his initial panic subsiding as the count of safe members grew with each moment. But men were still missing. "At this point, it felt like it had been awhile, and Kauri said to me that we were running out of time," Cox remembers. While he paused to confirm the headcount of those who'd been found, Kauri and Nydegger made their way farther down. Until this moment, the numbers on Nydegger's beacon had been bouncing back and forth. He was slightly below everyone else. "I'm just praying for a number," he recalls. "Suddenly, the beacon makes a beep and I'll never forget that number—forty-eight."

A signal had been picked up forty-eight meters below him. Nydegger and Kauri covered the terrain, closing in on single digits until they hit a spot revealing a hand and the barest image of a face—Jared Flitton. Kauri was at Jared's side in moments.

"I yelled frantically that he needed to find everyone else. . . . He began searching around me and no longer than a few seconds passed when he yelled to me that he had found someone else. I remember him saying, 'I found somebody, a glove, he's buried. I'm uncovering him now.'" After hearing the sound of frantic digging, Jared heard Kauri call over his radio to Gore: "I found them, one has no pulse, and he is not breathing. I'm starting CPR. Get down here as soon as you can."

Jared yelled: "Who is it?"

"Yellow pants, blue coat," the guide shouted.

"That's Jake!" Hearing Kauri give repeated chest compressions, Jared Flitton could not believe what was happening. In retrospect, the men deemed it a miracle that the two buried skiers were so close together and could be attended to quickly. "While he worked," Jared remembers, "he was begging for Jake to take a breath."

Nydegger rushed toward the frenzied fight to restore Jake Rabe's pulse and breath. "Nothing could prepare me for what I saw. It looked like he had gotten hit in the face with a baseball bat. . . . He was totally unconscious."

After being dragged hundreds of yards down Mount Mackie and rammed repeatedly into trees and rocks, Jake's face was swollen and severely cut. His eyes were bleeding, Nydegger recalls, and his tongue had been gouged by shards of ice and other debris. It reminded Nydegger of a dead motorcyclist he had seen when he was a boy. "I saw Jake and I thought he was dead." Feeling powerless, Nydegger watched Kauri "punching him in the chest, hard." The guide and Nydegger began in unison to urge Jake out of unconsciousness. "We kept telling him, 'Jake, breathe slow.'"

As the two fought to revive Rabe, Tyson Cox and Jason Bond approached, checking on Jared Flitton who seemed clear-headed to them. They surveyed Jake's situation. "My heart sank when I saw him," Cox recalls. "Kauri was trying to get him conscious. . . . Kauri kept saying, 'I need you to breathe for me, Jake. I need you to breathe.'"

Suddenly Rabe showed signs of life. He was sucking shallow breaths in a rasping, irregular beat. The men encouraged him. But Nydegger felt at that moment they couldn't depend alone on Rabe's will to live. "We're men of God," Nydegger proclaimed to the group. "Can we give him a priesthood blessing?" As a lifelong Mormon, Nydegger is a "holder of the priesthood." In special situations he can "do the priesthood blessing by a laying on of hands." So while Kauri continued to draw Jake back to consciousness, and the others stood witness, Nydegger placed his hands on Jake's head.

"I said to Jake, 'I command you to be healed. I command you to be healed.' I was almost loud and yelling." Nydegger thought of Jake's wife, Jill, whom he had met days earlier at the airport in Spokane. She had seemed like a calm person, the wife of an ER doctor accustomed to hearing about life and death struggles, who remained serene in the face of such stories.

"I kind of had a change of heart," he recalls, pausing with the prayer. When he continued, "it was peaceful and more calm. I said, 'Jake, you will be healed.'" Not one to exaggerate such a frenzied moment of religious fervor, Nydegger says he heard his own voice change in the blessing's final words. "It was like God's voice," he recalls.

Miraculously, as if responding to the blessing, Jake Rabe stirred, trying to answer questions. Jared Flitton remembers him struggling to respond to questions about where he was and why he was in so much pain. Nydegger thought again of Jake's wife and asked the man to say something about her.

"She is friggin' hot," Jake blurted.

"Everyone laughed when we heard this," Jared Flitton says. "That was the Jake we were looking for."

RADIO CALLS FILTERED THROUGHOUT THE Big Red Cats' operation as the first signs of the sunset inked the sky. Snowcats filled with other clients were ordered back to the staging area so those additional cats could be deployed to help with rescue operations. On the snow, guides towing tobaggans were dispatched to Mount Mackie to bring the injured down through cliffs and thick trees. As the sun dropped behind the Selkirks, the temperature began to fall.

If it seemed to the men that the rescue would now accelerate, instead the mountainous terrain slowed the pace of crews coming to get them. Soon darkness set in. And then it began to snow.

As relieved as some of the men felt about having survived, the danger wasn't over for them. Kauri told his group to switch their beacons back to "send" mode in case another avalanche broke loose and swept over them before the rescuers arrived. The small group helping Jake Rabe and Jared Flitton huddled together. Other members of the party were still spread out on the mountain above them. Iverson's leg had been broken, and they had used some broken ski poles to fashion a makeshift splint for him. Green remembers learning Iverson had been hurt but that a cry of "full count" had gone out over the mountain—a shout he had echoed. "As I finished that sentence, I started to cry," Green says.

While Green, Iverson, and three others awaited rescue, Mike Crowton was asked to give a blessing. "For fear of dislodging the cornice that Clint is sitting on, I don't dare slide out to place my hands on his head," Crowton recalls. "I fold my arms and bow my head and pronounce a blessing of comfort upon Clint. . . . From the instant that I uttered those words, a calming sense of peace washed over me."

At least three hours passed. Coats were offered to those most hurt. Nydegger laid his body close against Jake Rabe's as a guard against the injured man falling back into unconsciousness. Above the group with Iverson at the higher elevation, a small light eventually appeared from the guides with the first rescue toboggan. It was swiftly decided that it should be brought down for Jake Rabe and Jared Flitton, who were judged to have the most serious injuries. Jared "was shivering uncontrollably and beginning to feel terrible. I couldn't think straight and from what I could tell was beginning to go into hypothermic shock."

Those who could still click into their skis, and Nydegger into his snowboard, began to slowly edge their way down Mount Mackie. While the trees blocked their view, they could all hear the sound of snowcat engines humming on the road below. When they reached the two waiting cats, Kauri and the men loaded Jake Rabe and Jared Flitton into the first cat, as Nydegger and Brandon Flitton jumped in to accompany them.

"As we prepared to leave," recalls Jared Flitton, "Kauri stepped up into the cat and told me that I was going to be okay and kissed me on the forehead." Cox and Kauri watched the snowcat rumble its way along the dark road, its powerful headlights playing against the trees as it disappeared. The two men stayed to help with the remaining injured.

Cox recalls: "I hung out with Kauri for about two hours while we waited for the others. He kept repeating how shaken up he was. He said a few times how he had gone over the scenario hundreds of times and trained for something like that for over ten years. He repeatedly told me how well we all did and that it was a miracle everyone made it out alive."

Four guides, among them Big Red Cats owner Kieran Gaul, brought Iverson and the others down. Using cell phone flashlights, the Utah men picked their way down, having to lower Iverson over a cliff in a harness at one point. So steep was the terrain, the four guides had to stabilize the toboggan awaiting Iverson with each guide on a corner, using skis and snowboards as snow anchors.

Kauri and Cox watched the second group being brought down and loaded into the second snowcat. They began an hour-and-a-half ride back to Big Red Cats staging area, where the men found a massive gathering of police cars and fire trucks, paramedic trucks, and red and blue lights flashing everywhere. The scene resembled first responders at a plane crash.

Ben Rabe, Jake's brother, was told it was Kauri who had found Jake under the snow, dug him out, and performed CPR on him. "As soon as I saw him [Kauri], I went over to him and gave him a big hug and told him 'Thank you!' I asked him how he found them. He told me that he couldn't really explain it, but that he felt like he had been 'pushed or carried by some force' to where they were. He said he couldn't explain it or the feeling other than to say a 'higher power' guided him to them."

By this time, through social media and brief cell phone calls, the twelve men's wives had learned about the accident. After talking with family members from inside vehicles where heater fans were on high, emotions began to spill out. Some men were now just learning of the scope of injuries. Crowton spotted Kauri and hugged him. "My emotions start to get the best of me," he says, "so I walk away and just break down." Reuniting at Big Red Cats' base camp, Nydegger described to Crowton Jake's and Jared's injuries. "I can't handle it," Crowton recalls. "Again I break down, unable to control the flood of emotion."

Nydegger sat in a truck for more than an hour, the heat blasting. "Thinking about my family, I thought about each one of my five kids. I thought about my wife. I thought, 'I've got to be a better dad.' It was a pretty spiritual experience."

Those men who were only shaken up and had no serious injuries were taken to Big Red Cats' lodge, where they were fed a late dinner and debriefed about the accident. Meanwhile Jake and Jared were rushed to Kootenay Boundary Regional Hospital, about an hour away in the small town of Trail.

After the receiving doctors had assessed Jake and run CT scans and other tests, an overseeing physician reported he had a right clavicle and scapula fracture, right tibia/fibula fracture, several broken ribs, facial fractures and contusions, and hairline fractures to some vertebrae but an intact spinal cord. He also had a "brachial plexus injury," which was eventually determined to be a major nerve severed in his right arm. The next day, he was taken by ambulance to the Canadian border, where a second ambulance picked him up and took him to Sacred Heart Hospital in Spokane—the very facility where Jake Rabe worked as an emergency room doctor.

Jared Flitton received X-rays that showed he had not broken any bones, though one knee was "banged up" and a doctor recommended he see an orthopedist when he got home. "I guess you were just lucky and didn't hit anything on your way down in the avalanche," the doctor told him. "I almost laughed at that comment," Flitton recalls. "'Didn't hit anything.' I hit *everything*!"

After checking on Jake and Jared at the hospital in Trail, the other men went back to their hotel in Nelson, well after midnight, and tried to get some sleep. But for most it was elusive. "Every time I closed my eyes, I would see the slide—again and again," remembers Green. "I kept trying to ignore it but every time I started to doze off, I would wake up thinking I heard the beeping of my beacon. It was the worst night of sleep in my life. . . . It was a lonely, dark, and scary night."

Iverson ended up being taken to Sacred Heart for resetting his broken left leg. As he was being prepared for surgery, all the men who had driven back from British Columbia to Spokane the morning of January 20 showed up at his hospital room to give him a final blessing. "The power of the priesthood was real on the mountain," he says. "These guys were worthy and prepared long before the crisis. They called down the powers of Heaven."

OVER AND OVER, THE TWELVE men would play back their memories of January 19 on Mount Mackie. Maybe one could have pointed to human error or frailties that played into the accident, but they preferred to focus on their respect for Kauri and other guides and how they had handled things. "I couldn't say enough good things about Kauri and how well he responded to such a terrible situation," Cox says. "I thought [Jake] and the others were gone. But miraculously and by God's guiding hand and the help of the Spirit, Kauri was led directly to Jake and Jared in the exact moment he needed to find them."

"I owe Kauri my life," says Jake Rabe.

Frequently, the men reflect on what they insist was the providence of God. "I am so grateful for a God who is in the details," says Cox. "He was watching over all of us, and I know that is the only reason we all made it out alive." Of the experience, Ben Rabe says: "All of these guys are amazing, and I know that there is no way that we should have survived this avalanche. The only reason we did is because of the watchful eye of a loving and caring God."

Jared Flitton reflects on "the miracle that was the avalanche. Smacking into trees only to bounce off of them with no major injury. A hand perfectly positioned in front of my face to clear the snow. The familiar voice of a rescuer that was able to hear me call to him. An outstretched hand that grabbed me, seemingly pulling me to the surface, and the words 'I found you, you are going to be okay.'"

With the exception of Jake Rabe, all the men's injuries healed. After surgeries to repair his fractures, efforts were made to restore the use of his right hand, but extensive rehab brought only faint finger dexterity. A year after the avalanche, Jake was still not able to resume emergency room work, but his faith and willpower were working in his favor. "Many physicians . . . have frankly stated that I would never move my arm again. Well I am. The high that was evoked when I saw my pinky finger flex half a millimeter was indescribable. Sensation is absent, so I feel like a bloody Jedi master willing this foreign extremity to obey me."

Jake Rabe has had a lot of time to reflect on the gift of life, his family, particularly the help of his wife, Jill. "I have been given a unique opportunity to view and value my wife in a way that seems immeasurable. My love and appreciation for her is at a place that I don't think I could have come to, otherwise, in this life. Our relationship was sincerely happy before, but now it seems impenetrable. . . . I have the privilege of continuing to be a husband and father in mortality. That is the true gift."

Over time, the men have been able to find simpler celebrations of surviving the avalanche. When they decided to get together for a kind of reunion in January 2018, Jake Rabe wasn't able to come, but he sent a video. He told the men if he could move his right hand and fingers, "I would flip you guys off."

"One other thing," Chris Bond told the group, whom he now thinks of as true family, "I've been singing 'Baby You Got What I Need' for the last two months."

Says Ben Rabe: "I went to Canada with one of my brothers, but I feel like I came home with ten more."

13

SO YOU WANT TO BE A SKI-PORN STAR

In the late 1990s I became obsessed with a ski film called *Snow Drifters*. Set in locations all over the world, the film featured some of the best skiers of their time, raging in the backcountry, doing flips off of cliffs, or "hucking" as they called it, submarining through chest-deep snow, and drawing lines down impossibly steep couloirs. This kind of ski film fare was pretty typical of many such videos in the 1990s. But there was a sequence in *Snow Drifters* I couldn't take my mind and eyes off of. Set in a little mountain hamlet named La Grave in France's southeastern "department" of the Haute-Alpes, a noted "extreme" skier, Micah Black, had come to ski the dominant mountain in the valley, La Meije.

Known for its lethal steeps, La Meije isn't a ski resort. It's a 13,071-foot mountain you can ascend by way of a gondola system, which ultimately drops you onto the Girose Glacier. From there, assuming you don't vanish into a crevasse, you can descend whatever off-piste route you choose. There are no official trails, no ski patrol, no avalanche mitigation. To ski La Grave without either serious time on the mountain or an experienced guide is asking for trouble of the sort that almost certainly places you within the close reach of death. La Grave—the English word alone invokes anxiety—is where Doug Coombs, one of the most famous American skiers in history, met his end while trying to save a young ski protégé who had fallen over a cliff.

Snow Drifters, released in 1995, had been produced by Canadian James Angrove, who made a number of similar films under the banner of his company Real Action Pictures (RAP). I long ago loaned—and lost—my VHS copy of the film to a young family member, and when I tried to track it down online, using the company name, all that came up were rap or hip hop films, such as *8 Mile* and *Get Rich or Die Tryin*. Eventually, I located a website from which Angrove apparently still sells VHS copies of the film. His description of it brought back memories of my fixating on many sequences in the film: "Discretion is advised as this baby packs some controversial scenes of our skiers and snowboarders letting loose at night after near-death experiences. Don't buy this one for your kids!"

The thing that brought me back again and again to watching the La Grave sequences were the skiers' techniques on the near-perpendicular slopes of La Meije. So close up were the often slow-motion telephoto lens shots of Black and his companion, I could see how their knees were absorbing jump turns, how they were reaching their ski poles down the face as they initiated each turn. The mesmerizing skiing was all synced to a druglike Jimi Hendrix soundtrack. After watching the La Grave skiing scenes scores of times, I reached a conclusion: I had to go there.

My impulse (which I acted on) was significant because as the 1990s ended and the next century began, it ushered in a video technology revolution that shrunk cameras, like the early Sony Handycam, to a size that any teenager could use to produce his or her own personalized version of the films Angrove and others were making. Skiers and snowboarders invented all manner of rudimentary ways to fasten cameras to chest straps or the sides of helmets using duct tape. Or they had friends set up with tripods right in front of jumps or downslope to catch footage of onrushing skiers and boarders making sinuous turns and throwing up huge fans of snow behind them.

AS THE NEW CENTURY DAWNED, small-camera technology accelerated, with the hardware sizes getting smaller and the video quality getting sharper. Meanwhile, this technology was paralleled by advances in another new medium—the internet. By 2004, when a surfer named Nick Woodman launched a new miniature camera called the GoPro—aka the HERO cam—the World Wide Web was primed to receive an onslaught of close-ups. No backdrop was more ready for those shots than skiing's and snowboarding's backcountry. Suddenly, would-be filmmakers, equipped with a couple thousand dollars' worth of GoPros and some laptop video-editing software, could produce ski mountaineering films of a quality that far surpassed the films of the 1990s.

In their own vintage way, older, sometimes grainy films like *Snow Drifters*, or the trailblazing Greg Stump film *Blizzard of Ahhs,* still have a place in ski iconography. Some of the ski stars in those films, like Micah Black, remain classic pioneers who drew lines on mountains with enviable technique. Trevor Peterson, Eric Pehota, Doug Coombs, Kristen Ulmer, and others laid down turns that are still worth studying for their mastery. The ultimate effect of what came to be called "ski porn" (similar to its unruly cousin "surf porn") was that a younger generation began filming their own backcountry exploits and posting them on YouTube. They also tuned in to others' videos posted from all over the world. When they saw videos of locally accessible terrain, it made them want to rush out to those backcountry spots to ski their own lines and record those adventures for posting. It marked the modern expansion of the backcountry and spurred forth new generations of riders.

It made people like me want to travel by plane and train and van to a place like La Grave, France.

THE MODERN EVOLUTION OF SKI films began with a Californian named Dick Barrymore, who in the 1960s and 1970s, following the surf film lead of Bruce Brown's *Endless Summer,* produced a series of twenty-eight long-form films and shorts on ski bum adventures in the mountains. Beginning in 1960 with his first film, *Ski West, Young Man,* Barrymore stoked the country's obsession with California "cool" but did it through a tight focus on the skiing life in exotic mountain terrain. After his first film, Barrymore released such titles as *Some Like It Cold* (1962), *A Cool Breath of Fresh Air* (1966), and *Last of the Ski Bums* (1967). His final release was in 1997 with *The Golden Years of Ski Films.*

A close friend of world champion surfer Phil Edwards, Barrymore had strung together his films with hip narration and the dancelike gyrations of "hot dog" skiing. His efforts ran parallel to those of the man who a decade earlier had started on the mountaintop to become the P. T. Barnum of ski movies—Warren Miller. After the end of World War II, Miller had sensed that a party was gathering in the mountains around Sun Valley, Idaho. Setting up shop from a trailer in the ski area's parking lot, twenty-two-year-old Miller bought an eight-millimeter camera and sought to bring his native Hollywood sensibility to the ski slopes.

For the first few years he took his amateur footage and edited it into something he could show friends in Los Angeles and performed the narration in person. With his sometimes corny but endearing sense of humor, and an avuncular voice, Miller adopted the irreverent attitude of a beloved uncle narrating film footage as if from family ski vacations. Beginning with *Deep and Light* in 1950, which he made with a borrowed sixteen-millimeter camera, year after year Miller stitched together footage from America's nascent ski destinations. As the venues and audiences grew, so did the map of Miller's travels to Europe, including Scandinavia, and other exotic ski locations. Over time, his style of filmmaking and narration was more finely honed until it was identifiable as the "Miller style"—a formula of expert skiers in idyllic mountain terrain, with gasp-eliciting wipeouts and a jester's predictable quips. The montage became an annual fall pilgrimage for skiing families all over America.

By the 1960s, Miller was booking his screenings into more than a hundred cities. So successful was the franchise, he had to start recording his narration as he was unable to attend every showing. In his autobiography, *Freedom Found,* published in 2016, Miller said of his ski and snowboard films: "Looking back on what set my films apart, it was the emphasis on entertaining people, which means making them laugh." Capturing the magic of white worlds and the freedom of life in the mountains, Miller's audiences streamed from the theater with dreams of the winter season ahead.

THE FORMULA IN MILLER'S FILMS, and Barrymore's capture of the "cool" factor in skiing, had not escaped the attention of other younger, would-be ski movie makers. Most prominent among them was Greg Stump, who made the ski culture–shattering film *Blizzard of Ahhs.* Later on, the Jones brothers—Steve, Todd, and the famed snowboarder Jeremy—founded the ski film company Teton Gravity Research in Jackson Hole, Wyoming, a company that launched a new wave of rock and danger-infused videos.

Stump, who grew up ski racing in Maine, met Barrymore in his late teens, not long after Stump had won the Junior US National Championships. The elder ski filmmaker took Stump under his wing and offered him a chance to help make Barrymore's 1979 movie *Vagabond Skiers.* The experience, which led nineteen-year-old Stump to follow Barrymore to shoot windsurfing in Hawaii and skiing in Australia and New Zealand, left an indelible mark. Stump had learned techniques that he could use to put his own mark on films for a younger generation than the ones Barrymore and Miller catered to.

But before he became an auteur of the slopes, Stump realized he had to broaden his mindset and go to college. The decision played a critical role in the making of *Blizzard of Ahhs,* not necessarily because of the classes he took at the University of Southern Maine but because of his part-time gig as a DJ at a nearby radio station. Based on his own competitive skiing prowess, Stump secured a role in a 1983 Miller film, *Ski Time.* Absorbing more about moviemaking technique from the experience of being around Miller and his film crew, Stump began to formulate his own narrative approach to making ski films. Rather than being overly reliant on his voice telling the viewer what was happening on the screen, Stump drilled down into his DJ immersion and found a powerful voice in rock and roll music.

After his first four films, Stump had managed to cobble together enough sponsorship money from the K2 and Salomon ski companies to pay his own skiers and foot an ambitious travel budget. Scott Schmidt, a stand-out ski celebrity of the 1980s who had starred in Miller films, was partly persuaded by Stump to sign on because of the prominent sponsors involved but mainly because Stump promised him extended filming in Chamonix, France. Schmidt was among a generation of skiers who had begun to blend mountain climbing techniques—using ropes, harnesses, and crampons—with assertive, high-angle backcountry skiing, a departure from Miller's resort-focused skiing.

If Stump had hit on rock music as a key, calculated element of *Blizzard,* his next break was unplanned and initially seemed a setback to the production. Shortly after arriving in Chamonix, one of his on-camera stars was injured, and Stump was forced to enlist a substitute. In an earlier movie, he had filmed Nevada skier Glen Plake, who was dynamic on the mountain, although he projected an outlaw-like, radical persona. Plake sported a Mohawk haircut and a swagger that didn't dovetail with Stump's increasingly earnest business attitude toward his productions.

"Even though Glen was great on film," Stump recalled in a published interview, "I was very turned off by him. He was a heavy drinking, substance user into punk rock. . . . His endless cackle and chatter drove me nuts." But once he had Plake on-site in Chamonix, Stump saw through his camera lens that the unlikely pairing of Schmidt, an austere ski mountaineer, and Plake, an extraordinary skier with an unlimited bag of tricks on snow, was a masterstroke. Paired with heart-thumping music, what Schmidt and Plake accomplished in *Blizzard of Ahhs* was unlike anything audiences had ever seen in ski films.

Twenty years later, Stump remarked that "people have been congratulating me on pairing the polar opposites of Schmidt and Plake. . . . I was obviously wrong about Glen . . . and he helped my career immensely." Schmidt went on to star in more ski films, and Plake, who curtailed his partying to focus on his skiing, was inducted into the US Ski Hall of Fame in 2010. The release of *Blizzard of Ahhs* in 1988 was transformative in ski movie making. NBC and sportscaster Bryant Gumbel invited the two stars to appear on the *Today* program. Schmidt showed up as the clean-cut, golden boy in a ski sweater—as if Robert Redford had stepped from a scene in *Downhill Racer*—while Plake appeared in the studio dressed in an American flag suit, his skull topped by an immense, carefully sculpted Mohawk.

National exposure for the two seemingly contradictory personalities, and sequences from the movie of their "extreme" skiing, caught the attention of an emerging generation of adrenaline-sports enthusiasts—skiers, snowboarders, short-board surfers, skateboarders—who could identify with Schmidt and Plake. Stump's film received, unlike most previous ski films, and even Miller's annual roadshows, mainstream attention. In a September 2012 interview with the *Denver Post,* Stump said that when he travels and encounters skiers and snowboarders, they invariably say, "'Your movie changed my life.' I get that a lot. It is the greatest compliment ever."

STUMP'S WORK, AND BEFORE IT Miller's and Barrymore's, changed the lives of the Jones brothers, Steve, Todd, and Jeremy. While there were other start-up ski and snowboard film companies in the 1990s, many inspired by Stump's work, the Joneses' Teton Gravity Research (TGR) became the new gold standard for contemporary ski and snowboard films. But the Jones brothers undertook their early work the hard way.

From the inception of ski films in the 1960s through the 1990s, there was a significant barrier to entry for turning out ski film productions—the cost of such endeavors. Camera gear was expensive, as was film and the processing and editing of film. Few dared to venture into a business that was expensive, time consuming, full of travel logistics, weather problems, and the potential for stars and support people to sustain serious injury in steep mountain terrain.

In 1999, TGR released its first full-length film, *The Continuum.* Funded by money the Jones brothers made from commercial fishing in Alaska, *The Continuum* took a few cues from Stump—a score of rock music—and brought to the screen a point of view from the open doors of helicopters hovering around

Alaskan peaks that seemed just feet shy of sheer vertical. What TGR and the Jones brothers documented was essentially a rebellious tribe of mountain athletes who had rejected the controlled environment of the ski resort to find a new kind of freedom in the wilderness. They were ski hipsters who were literally dropping out so they could drop in to first-descent runs in places like Alaska's Chugach Range.

The Jones brothers, whose early films featured their own dramatic skiing and that of other luminaries like Doug Coombs and former US Ski Team member Jeremy Nobis, had hired off-duty oil rig helicopters to ferry them into big mountain ranges. Left marooned on virgin snow peaks while the helicopters circled with cameras running, solo skiers took the only escape route at hand—an elevator drop off the peak down a three-thousand-foot spine where the risk was "you fall, you die." "Everything was expensive," Todd Jones said of TGR's beginnings. "A hundred dollars a minute to shoot film. Editing was thirty thousand dollars. . . . It was an exclusive club and you had to really want it."

There was another big distinction that set TGR apart, recalls Jim Conway, who headed up TGR's safety program in the early going. Their skiers and snowboarders were elite athletes engaging in feats at the outside of the performance envelope, not joy-riding performers on holiday. Conway saw *Blizzard of Ahhs* and was impressed with the new model it presented, but he remarked to skier friends: "We do that shit everyday." He says, "TGR was always pushing what human beings could do in the mountains," noting that the Jones brothers had briefly skied for Warren Miller but found his films too limiting. "The Jones brothers got so frustrated they got mad and went commercial fishing and then bought camera gear. Their premise from the start was to express the athletes' vision."

"We didn't hit it off so well with Warren Miller," Todd Jones told me. "We said, 'Screw it' and bought a camera." He and his brothers had glimpsed the athleticism Schmidt and Plake shared in *Blizzard of Ahhs*. "We wanted to bring that back and acknowledge the athletes because what they were doing was super amazing stuff."

But in Alaska's often alien, unpredictable terrain, it became apparent that safety would have to be paramount, or the Joneses would end up filming the ski equivalent of a snuff film. Conway said the TGR team started doing safety briefings in the mornings before filming. When the weather was marginal, they chose to sit it out. They once even had a safety stand-down, Conway said, a suspension to filming, when "two close call avalanche incidents" occurred. "No one was caught," he said, "but avalanches were triggered that could have had serious consequences. Basically, we lucked out and we don't want to rely on luck."

It's to TGR's credit that after two decades of productions among the world's most vertical and dangerous mountain descents, with athletes jumping onto those peaks out of buzzing choppers, there have been no fatalities. Broken bones, torn knee ligaments, and lots of cuts and bruises have piled up. But the athletes' conditioning, the TGR safety protocols, and overall regard for risk management—particularly in avalanche-prone terrain—has led to stunning film sequences and enhanced the athletes' standing. If there is a standout example of how close an athlete skier can come to serious injury or death and yet, through experience and physical prowess, walk away largely unhurt, it's pro skier Angel Collinson's fall in a run in Alaska in 2016, captured on video by a TGR crew.

Utah native Collinson, a two-time world champion on the Freeskiing World Tour, has been a featured TGR skier for a number of their films. One mouth-dropping run after another, she nails her lines in the film *Almost Ablaze*. A YouTube segment of her runs for TGR is titled "Angel Collinson Annihilates Alaska: The Rowdiest Women's Skiing Segment To Date?" In truth, it should have been called: "The Rowdiest Skiing Segment to Date."

Collinson's risky Alaskan descents brushed a fine line in spring 2016 when she dropped into a couloir on a run she hadn't even intended to be in a TGR movie, although the team was filming. She dropped in, she told me, because she wasn't feeling her usual self, and "I felt like I needed to push through." "Pushing through" ended up meaning struggling to stay alive. Posted on YouTube, the video is headlined "What Was Angel Collinson Thinking While She Tumbled Down This Massive Couloir?" It begins with a helmet-mounted camera angle directed at Collinson's skis, and the radioed words of a TGR filming team member saying, "Three, two, one. Go! Go! Go!"

Then Collinson kicks in. For about thirty-five turns the viewer sees the tips of Collinson's skis knifing through snow and ice, her pole plants strong and confident. She's acing it. Then, about halfway down the couloir, she cuts to her left, preparing to make a turn back to the right, and her skis hang up in dense, shrapnel-like snow. She begins to lose the edge that was holding her skis in the turn. She goes down, her left arm with her ski pole trying to dig into the snow like a flesh-and-bone pick axe, a self-arrest. But she's adrift, at the mercy of gravity now.

The camera POV changes from Collinson's point of view to an angle from the valley below. She falls end over end, "tomahawking" skiers call it. Watching it, you have only one possible reaction: you hold your breath. The tomahawking grows worse and worse. She slides over a faint ridgeline that plummets steeper, and she gains speed, a human boulder waiting to slam into something. Her voice, picked up by a microphone, is a series of deep breaths and uttered animal sounds, like desperate pleas for pain to stop.

Finally, as the slope angle moderates a little, she slides into a runout, flatter snow on which her speed decreases. Then she halts. Seemingly lifeless, her form a bundle on the snow, seconds ticking away, you wait for some utterance from Collinson, a movement. Anything.

There's some chatter over the radio, then Collinson's surprisingly vibrant voice: "I'm okay, I'm okay." It's a tribute to her skills and her coolness under dire circumstances. It's also a tribute to TGR's determination to learn from accidents, since Collinson was asked by them to narrate, in person, the fall to a group of pro skiers and snowboarders as a learning experience.

If, over the years, fine-tuning TGR's safety and production operations in the mountains and the individual films' narratives has been a big part of what distinguishes the company's direction, other things evolved in tandem—the miniaturization and plummeting cost of digital camera technology and the internet as a video-hosting medium. "We built a website out of the gate," recalls Todd Jones. "I immediately was struck that I didn't have to fly down to L.A. to the big boys club and ask for distribution."

Within a relatively short period of time in the early to mid-2000s, TGR had exploded into a brand that not only made ski and snowboard films, it also began to feature some surfing video on its website, and had a clothing line and chat and blogs that would be the envy of any social media manager. "A lot of people didn't embrace that, but we did," Todd Jones says. "We saw a bigger world and we started transitioning into a media company. We weren't threatened by it." While TGR had so firmly established itself in the snow sports world and in the adrenaline-sports video realm, the financial dynamics that swiftly altered the production equation—cheap cameras and cheaper distribution—brought some smaller producers down.

NOAH HOWELL, A PROFESSIONAL BACKCOUNTRY skier and guide from Utah, in 2005 launched a small ski film company with his brother, Jonah. It was called Powderwhore Productions because that was what Howell and his ski buddies called themselves. Their intention was to show more of the mountains' serene beauty and what it took for skiers to place themselves in isolated backcountry terrain. "We wanted to share more of the real experience," Noah Howell says, "like how did you get there. You didn't just magically get to the top of the mountain."

The Howell brothers opened their first film in five cities—Salt Lake, Denver, Seattle, Lake Tahoe, and Portland, Oregon. "It was small crowds," he recalls, "but they were on fire." By their seventh film, the Howell brothers were on a forty-city tour, premiering to audiences of a hundred to five hundred. But Noah was beginning to see the end of the enterprise as vast numbers of videos from GoPro cameras and other digital technology were swamping online venues like YouTube and Vimeo. "There was so much footage online," he says, "and no one wanted to pay at that point."

Early on, in Powderwhore Productions' growth as a business, Noah Howell said it would cost them a dollar to reproduce a DVD from a finished film, which they could then sell for twenty-nine dollars. By the end, in 2015, when their tenth film, *Something Else,* debuted, Howell said the financial equation was down to little more than breakeven, and the demands of travel and administrative tasks had worn them down. But he doesn't regret the decade's roller-coaster ride, and feels that he and his brother produced a series of high-quality films with a different tenor. "I'm really proud of what we did," Howell says. "No one ever died. I never had to dig anyone out of an avalanche.... We showed a full balance of what the backcountry experience is."

Todd Jones says that in the early 2000s, ten to twelve full-length ski and snowboard films would launch every fall in anticipation of winter. Today that's down to two to three. "It's weeded things out," he says of digital video and the internet. "I think it's a good era. I feel privileged to have lived in this era." Ultimately, TGR is in the storytelling business; it just happens the story lines are usually in the mountains, although increasingly they are moving into other areas, such as a recent feature-length film about the life of the late pro surfer Andy Irons.

TGR leverages all manner of media technology to generate and distribute their work today—mini cameras, drones, websites, Instagram, blogs. Such proliferation of video, still images, and the overall vibe of contemporary boarding culture—dress, trends, language, music—has undoubtedly had an influence on the growth of backcountry riding. But it's also being used in new ways to highlight changes in resort skiing, where "trick parks" have been built, snowcat tours into side-country have expanded, and more gates out of the resort into the backcountry have opened.

Recently, TGR did a series of mini ski and snowboard films for eighteen resorts in twenty-one days, Todd Jones says. All the resorts are linked through the ICON ski pass, which admits holders to any of the resorts. The TGR team would shoot all day, edit at night, and send out the films the next morning. "It was a really cool way of doing it on the fly," he says. "Twenty-one days, in and out, done. It reached a lot of people. . . . Social media isn't boring. That was a twenty-one day story in real time."

TGR released its twenty-third film, *Far Out,* in fall 2018, to be followed in 2019 by *Winterland.* Todd's brother Jeremy Jones, one of the top snowboarders in the world, was featured in 2018 and Todd's eleven-year-old son, Kai, made an appearance. "It's a really trippy thing," Todd says about his son's inaugural appearance nearly twenty years after *The Continuum.* "He didn't exist in thought when we had the TGR concept." Todd's younger son, who is nine, isn't far from the family business. He's "into cameras and drones."

IN THE SPRING OF 1997, I changed jobs and took a two-week break between them. For some time I had been thinking about the footage in *Snow Drifters* that portrayed La Grave and the ski runs on La Meije. Searching ski magazines, I found the name of a company offering ski guiding in La Grave. The American guide who ran the company was Gary Ashurst.

When I arrived in La Grave, I checked into La Chaumine, the small, Austrian-style lodge that Ashurst worked out of. A deck off the upper floor of the lodge faced La Meije, the main mountain. The trippy Jimi Hendrix music I'd heard on the *Snow Drifters* soundtrack played in the small bar. It was as if I had parachuted, via a time machine, into the world of the ski film.

Later though, perched on the La Chaumine deck as the sun was going down, I grasped the menacing reality of the difference between studying a mountain in a film and standing before it, knowing I would be skiing it the next morning. It was like landing on a new planet and realizing I was expected to explore its alien reaches with my only previous experience being some other hikes in distant friendly foothills.

While I'd checked in, a man behind the desk had handed me a key dangling from a small wooden tablet. "You're in Arnica," he instructed.

"What is it? Arnica?" I'd asked.

"A flower," he'd responded, pointing out what was painted on the wooden key ring.

I had revealed blatant ignorance of basic alpine flora. I may as well have confirmed myself to be completely out of my league, skiing in a place like La Grave.

Soon enough, I met Ashurst. A contemplative, compact native Californian from Sun Valley, Idaho, he had moved to La Grave in the early 1990s after skiing there on the advice of a friend. The fact that I had traveled there based on ten minutes of video came to light at some point. I couldn't tell if Ashurst thought that was adventurous or absurdly foolhardy. He gave me the benefit of the doubt, and as the week wore on he gave me ample credit for my passion for skiing, even if my abilities fell short.

People came to La Meije, Ashurst recalls, and said, "Oh my God, this is a different part of the sport. An adventure." My week there was certainly that. With no marked runs and no grooming—"off piste" as Europeans call it—the mountain was tough, steep, and, after about the thirtieth long traverse across a slope, turned my calves to jelly. La Meije was the pure "adventure" Ashurst's clients had deemed it.

We skied with avalanche beacons, shovels, and probe poles, my first experience with such gear. There were powder runs like I'd never imagined. One morning, at a nearby ski resort, we strapped our skis to our backpacks, hiked up to a peak, and dropped over the backside into the out-of-bounds. Descending on a pair of K2 Explorers, one of a new breed of backcountry skis, I found the sun had turned the slope to corn snow, something I'd heard about but never touched. Off the tails of my skis, the snow spun like beads of glass. The snow received me, not like someone unschooled in mountain flora, but as a devoted, humble pilgrim who deserved the mountain's blessing.

Twenty-one years after those mornings on La Meije, I reconnected with Ashurst. By that time he had relocated to Ketchum, Idaho. Now a cabinet maker and woodworker, Ashurst lives on a treed, flower-graced piece of property that hugs the Big Wood River. He has a tidy home, a Zen-like guest cottage, and a shop filled with enough tools to build out a western town set for an HBO relaunch of *Deadwood*. In some ways, his more than fifteen years of guiding in La Grave—with American skiers and snowboarders being the lion's share of his clients—stands as the human equivalent of a ski film's exposing remote, exotic terrain to the masses. During that time, La Grave went from a whisper destination, muttered in wax-dappled tuning bays of ski shops, to a skier's lifetime bucket list.

Ashurst saw just about every ski filmmaker of note make a pilgrimage to La Grave with a team of celebrity skiers in tow. Warren Miller, Greg Stump, James Angrove, and the TGR crew all came and took some direction from Ashurst. *Powder* magazine published one of the first articles about the town and mountain. Coverage in *Men's Journal, Snow Country,* and even in some advertising, such as for Marmot, turned up the wattage on La Grave. "I was getting a lot of press," Ashurst recalls, "because I was the only American certified guide. . . . The world came and skied with me."

Even with the media coverage, La Grave remained—and remains to this day—a place that is both alluring and foreboding. The 2006 death of Doug Coombs there, when he was arguably at the pinnacle of his ski mountaineering powers, provides a subtext to almost anyone's assessment of a trip there. Coombs was the confidence wizard. Based on that talent, Coombs and his wife moved their "Steep Skiing" camps from Wyoming and Alaska to La Grave. "Doug was amazing," Ashurst recalls, "an incredible person. Just a gas to ski with. He was the best I'd ever seen in terms of being a confidence builder."

There was also a downside Ashurst experienced around the fringes of his clients. Some showed up seemingly unaware they were under the influence of what was tantamount to a death wish. "I didn't just take anybody with a heartbeat," Ashurst says. "Sometimes I had to decide to cut people out and send them to the bottom." But for those with solid ski skills, and a willingness to embrace the adventure La Meije offered, the time with Ashurst could be memorable beyond imagination. He was quick to counsel his clients about the things La Meije held in store for the unwary. Sometimes he took them down a couloir that required them to wear a climbing harness and be lowered into the top of the chute. Clients often expressed mild protests just before going over the edge.

Ashurst had a stock response to the roped-up skiers about to negotiate a forty-five degree couloir: "It's the safest thing you'll do all day."

14

GEARHEADS: INVENTING AIR

I am standing in a hallway of the headquarters of Black Diamond, a ski and mountaineering gear company based in Salt Lake City, Utah. Jon Coppi, a quality engineer at Black Diamond, has just helped me sling a black backpack onto my shoulders. Stitched into the side of the backpack is its design name: Halo 28.

Coppi directs me to a short aluminum wand dangling from the left shoulder strap. At its end is a red cap and button. "Just press that button for a couple seconds," he instructs. With my left hand, I depress the button, cautious not to do more than Coppi has advised. From just beneath the rear of my skull a sound explodes, as if created by a jet-propelled vacuum cleaner, spookily loud, technologically borne. It's the sound of a new machine, announcing its genesis: you don't exactly wear *it,* it wears *you.* The *whoosh,* which lasts a little longer than my two seconds of depressing the button, shares something with James Bond jetpack fantasies. But this invention, a Black Diamond JetForce avalanche survival pack, has one vital mission: to help skiers and snowboarders survive burial by an avalanche.

I almost expect the next second to reveal a larger, even more profound detonation around my neck and top of my spine. But all is calm.

"That's just the button and sound that let's you know the JetForce pack is activated and ready to go," Coppi explains. "That just armed the system. What you heard was the fan blade in reverse at fifty thousand revolutions per minute." A tiny blinking green light on the wand indicates the pack is armed. A single blue light tells me I have enough battery power remaining for one deployment of the avalanche protection system. Had the battery been fully charged, four blue lights would have appeared on the wand, confirming enough power for four triggers of the system. Having the pack on gave me a strange, unexpected sense of security, a little boost of confidence, as if I was somewhat insulated from danger.

Lingering in an office hallway, near the entrance to Black Diamond, I notice a plaque commemorating the company's twenty-fifth anniversary. It reads: "We will always champion the vision, guts and drive required to forge a bold future for our sports, our tribe and the wildlands that define us." Black Diamond's employees and customers all over the world constitute this tribe. Walking through the vast design and testing facility offers a revealing look into this world. Throughout the headquarters, rooms are filled with young people, many of them looking like ski or snowboard bums, but clearly bums holding mechanical engineering degrees. The discipline lends itself well to people who want to design and improve gear that is used to have fun and excel in the mountains. It's also a deadly serious business that inspires a Zen-like devotion to the details.

Coppi leads me through the offices, past walls covered in large posters and black-and-white blown-up photos of rock climbers and skiers on vertical mountain pitches. "All these pictures are of our employees or Black Diamond team members and ambassadors," he says. One is of Tommy Caldwell on Yosemite's Dawn Wall, an El Capitan route that Caldwell and partner Kevin Jorgeson climbed in January 2015, a nineteen-day achievement that marked the first free ascent of the route.

In one room a glass-topped case holds scores of carabiners and other small climbing devices, all of them twisted like pretzels and snapped in two. "This gives you a pretty good idea of failure modes," Coppi says of the grotesquely mangled metal. Designing contemporary climbing and ski gear is only the start for these engineers. Testing the gear before it's brought to market is a critical phase. "Anytime we develop PPE—Personal Protective Equipment—a big chunk of our time is spent testing," says Coppi. "People are putting their lives in our hands."

All through the offices are custom-built test machines, bizarre-looking contraptions that open and close buckles and carabiners, bang ice axes against simulated granite, repeating the motion hundreds of thousands of times until the metal fatigues and on, say, oscillation number 63,110 the device snaps. Test machines record the results and evaluate whether the tool is within acceptable parameters. There are abrasion testers for ski jackets and pants, backpacks, and other soft goods, all ripped apart in the interest of better durability.

There are cold and heat chambers where gear—tents, avalanche transceivers, ski helmets, and other products—can be placed and subjected to extreme temperatures of minus 85 degrees Fahrenheit, then plunged into a furnace's near incinerating heat of 350 degrees. A salt-spray crucible blasts metal products and fabrics to see how they withstand corrosion accelerated so intensely that the materials age, in days or weeks, to degrees that would normally occur over many years.

These veritable torture chambers are for gear meant to be employed by customers taking the stuff to the outer limits of human endeavor. But those extremes are not even close to the breaking points the gear has suffered at the hands of Black Diamond's engineers. "We can take that data and feed it back to designers," Coppi told me, "to make things lighter, faster, stronger."

Although Black Diamond celebrated its twenty-fifth anniversary in 2014, the company's roots go back to the late 1950s, when it was founded in Ventura, California, as Chouinard Equipment Company. Its namesake, Yvon Chouinard, was in those days a young rock climber, surfer, and falconer. Often living out of a tent or in his car among other young climbers in Yosemite's infamous Camp 4, Chouinard started forging steel pitons he sold out of his car's trunk to other climbers. Eventually he expanded into climbing harnesses and a limited clothing line that began with rugby jerseys he had found in Wales.

Chouinard Equipment grew out of postwar California, where small beach towns were populated with aircraft engineers whose pockets were stuffed with cash from designing 1940s warplanes. Living in California's Edenic 1950s, they wanted a lifestyle of fun in the sun, going fast on Pacific Ocean waves and mountain faces slick with Sierra snows. So the engineers went into their garages and invented, using war-era technology—fiberglass and aluminum—new surfboard designs and skis, ditching balsa and other woods that had been used for decades in the sports. Today's Black Diamond employees are the spiritual descendants of those early innovators who wanted to create more ways to have fun in nature, giving rise to California's legendary surf and ski cultures.

Chouinard was one of those pioneers. In its early days, Chouinard Equipment won a small but devoted customer base that prized its reputation for unsparing quality. But because of the inherent dangers in climbing, product liability lawsuits, some without merit, became a distraction and expense that Chouinard eventually refused to tolerate. In early 1989 he placed the company into bankruptcy, and on December 1 of that year a group of former employees bought its assets and moved the operation to Salt Lake City.

Chouinard went on to found Patagonia, which grew into a venerated global enterprise that now enjoys one of the most respected reputations in the world for both its products and environmental activism. At the original site of Patagonia's Ventura headquarters, Chouinard's old blacksmithing forge, which he used for early pitons, still occupies a place of honor.

Renamed Black Diamond Equipment after its move to Salt Lake City, the firm grew quickly as so many of its customers were living and climbing in the Wasatch Mountains. In May 2010 the company was taken public and began an acquisition strategy to grow its offerings. Today it has annual sales of nearly two hundred million dollars, but it retains the feel of a closely held company where its employees display daily pride in working there: almost no one walks through its halls who isn't wearing a Black Diamond T-shirt, fleece pullover, or down vest. The shared sense of purpose seems to extend to every ritual.

In an engineering test room, five of Coppi's colleagues huddle over a pizza box, taking a late lunch. "Hey, save one of those for me," Coppi says, in a teasing tone. On a wall a poster bears the mantra: "To Do List: Drink Coffee. Fix Everything. Lunch. Drink More Coffee."

STILL WEARING THE HALO 28 JetForce avalanche backpack, I walk with Coppi through the Black Diamond facilities for nearly an hour. At Black Diamond, people are always walking around with mountaineering equipment under design, contorted from a gruesome testing regime, or being remodeled in a way that invites contemplation. Ruminations about gear—a mindset that the company approaches more like meditation—is at the heart of Black Diamond's innovation process. There may be no product that exemplifies this more than the JetForce avalanche survival backpack.

Coppi introduces me to senior engineer Pete Gompert, head of Black Diamond's ski design operation. A trim, bearded athlete, he is one of the few people not wearing a company logo on his T-shirt. On his shirt, the image of a mischievously smiling Abe Lincoln masked as Batman stared back at me. Before joining Black Diamond, Gompert worked for Autoliv, one of the largest automotive safety device suppliers in the world, providing airbags, seat belts, and steering wheels designed to absorb crash impact. The company boasts that every year its products save nearly thirty thousand lives. At Autoliv

Gompert became familiar with automotive airbags that use explosives to deploy, an option he knew was not applicable to a skier's device. He says he begged Black Diamond for a job until his persistence wore them out.

Gompert and Coppi worked together on the early design and evolution of the JetForce backpack, work that laid the foundation for what became a team of ten design specialists. While the project was long, tedious, and led them down some dead ends, the pair took on roughly defined roles. Gompert did a lot of the early engineering analysis, while Coppi assumed the semi-volunteer job of test pilot. They were a little like Dyson trying to invent a revolutionary vacuum cleaner—one without a bag and with a swivel head that bore more resemblance to a tricycle than a rolling suitcase. Only Coppi and Gompert planned to attach their contraption to a human being's back in the interest of saving lives. Dyson had only been bent on collecting more dust.

Black Diamond was not the first company to design an avalanche survival pack. At least a couple had come out with packs fueled by compressed gas canisters; when a ripcord was pulled, these packs inflated a large "pillow" around the skier's upper back and head to keep him or her on top of the whitewater-like boil of hurtling avalanche debris. The basic theory of avalanche backpacks was this: a proven law of physics called "granular convection" governs that larger objects caught in a swirling sea of smaller objects will be pushed to the top. Such forces can be demonstrated, Gompert told me, by taking a bowl of peanuts and placing a much larger nut, like a Brazil nut, in the bowl but covered up. Shaking the bowl will prompt the Brazil nut to peek its way through the smaller peanuts until it sits on top of them. Such was the theory for humans made larger by the backpack airbag being forced above loose snow.

Gompert's bosses had taken note of the devices and asked him to create one for Black Diamond. As he studied the packs, he felt there were serious flaws in them. Mainly, it was the gas canister ignition that ballooned the pillow. Only containing a single inflation, the gas canisters had to be professionally refilled after each usage. And the canister packs could not be taken on airplanes for fear of accidental ignition due to changes in air pressure as planes climbed to higher altitudes. Gompert decided he would start from square one. Early on, he thought the answer might be in some sort of chemical mix with a catalyst that would give it explosive properties—a kind of quick drip, controlled, small bang theory that would propel an airbag protectively around a skier's head. For a while, he had chemical containers scattered under his desk, but he became concerned something might migrate from one bottle to another and set off an explosion. He soon realized it was not exactly a human resources director's delight.

The real problem with chemicals, Gompert deduced, is that they are messy and over time have a tendency to corrode containers and leak. The lack of appeal to backcountry skiers was also an issue: the vision of plunging down a steep, snowy couloir with a charge of chemical dynamite strapped to your back is hardly a confidence builder.

Gompert felt a little like he was Major Tom in David Bowie's "Space Oddity." He let his mind roam through a kind of inventor's periodic table of friendly metals, gases, and pairings of those raw elements. At the most basic, what he needed was air bolting into a bag, like a helium tank filling a birthday balloon but far faster. He needed a system to control it so the user was master of the combustion, rather than being a subject to the whims of an ignition point that could take on its own fiery life. His thoughts went to his time at Autoliv and the airbag technology he'd seen there, but he rejected explosives. He liked electricity because you can decisively switch it on and off. There, he thought, was the master operator model. And he liked the notion of keeping the air in the bag clean of contaminants; it met Black Diamond's mission of "leave no trace" in the backcountry.

Gompert searched for his own blue flash of innovative spark. And then he looked at a computer. Not at the screen, or the keyboard, or the mouse. Rather, he thought about the back of a typical desktop. Every workstation has a little round vent on the rear of its boxy, metal enclosure. Behind that vent was a fan. Holding that image in his mind, Gompert made his way to Black Diamond's IT department. Did they have any spare desktop cooling fans? Finding one available, he took it and went on another search for a spare plastic office garbage bag. Carefully sealing the output vent of the computer fan to the garbage bag, he plugged in the fan. As the little blades whirred to life, he timed the inflating of the bag.

In just under fifteen seconds, he saw the bag was a belly of air, an imperfect life ring for cradling a skier's head and shoulders and buoying the skier to the surface of an avalanche. It was ugly, without any of Black Diamond's elegant craftsmanship, but Gompert felt he could make it beautiful. As he refined his concept, Gompert knew that speed was critical when inflating the bag. An avalanche materializes like a car crash, hence the need for a skier's equivalent of a driver's airbag. But most avalanches don't last more than forty-five seconds, and many last far less. He sought an eruptive burst of air that would fill a bag so quickly and rigidly it was like a tiny dirigible fixed to the skier's shoulders and head.

The computer fan had worked. But it didn't have the explosive power Gompert needed. He asked around Black Diamond if anyone knew about electric motors used in Radio Control airplanes. Based on some feedback, he searched the internet for the smallest, fastest fan he could find. He needed a durable fan that was part wind generator and part bomb. He finally found it in an unlikely place—a radio-controlled airplane company's website, Horizon Hobby, that sold a miniature, motorized fan for model jets that can fly over two hundred miles per hour. The motor's part was designated the "E-flite BL15 ducted brushless." Its name was Outrunner. When the motor was switched on, its blades spun at 70,000 RPM—1,333 revolutions per second. Fast enough to maybe even outrun an avalanche.

With the rocket-like fan motor, Gompert had found the beginning to a true inventor's trail. His mind jumped to the bag or "balloon" the motor would have to inflate. It needed to be a tough material that could handle the instantaneous impact of air gushing into it until it was hard as a drum head. It had to be highly abrasion resistant and virtually impenetrable to all but the most switchblade-like lacerations. Rock edges and tree branches—all dangerous avalanche jetsam—would have to have no chance of slicing into the bag.

The beauty in that challenge, Gompert knew, was that Black Diamond's designers had extensive experience with highly durable fabrics like one called Cordura. Certainly they could come up with something suitable. They had in-house machines for torturing those fabrics with ballistics tests. Ultimately the designers came up with a custom in-house twisted yarn material made for more stretch and tear resistance. They coated it on the inside with silicone to make it airtight and stiff, so it would be more resistant to things like tree branches, allowing them to slide off instead of puncturing the material. Gompert was moving closer to his avalanche rescue device.

But unknowns remained. Black Diamond's designers were mostly mechanical engineers, experts at working with steel, aluminum, titanium, and other exotic metals. They build things for mountain climbers like miniature suspension bridges such as their Camalot C-4s. Made of wedges of metal, short cables, and springs, cams are inserted into granite cracks on mountain faces.

When positioned just right, they can be pulled so they rotate and lock firmly inside cracks to provide anchors for climbing slings and ropes. Ingenious metal devices for mountain climbing are one thing, but using electricity to control a jet motor fan so a fabric halo could be blown up around a human's skull was another thing altogether. Like all Black Diamond designers, Gompert depended on exhaustive evaluation and spot-on precision to arrive at solutions. That's how the best backcountry skiers think. But sometimes luck falls like unexpected snowflakes. And Gompert got lucky.

IN 2012, BLACK DIAMOND BOUGHT PIEPS, an Austrian manufacturer of avalanche beacons and other gear. Black Diamond had sold avalanche transceivers before this, but they had never designed and built one of their own. PIEPS came with a known and highly trusted product line. And it came with something else—a team of electrical engineers.

The avalanche backpack project suddenly took on a priority and velocity Gompert's own tenacious efforts had lacked. Black Diamond assembled a team—backpack design experts, fabrication and materials experts, and PIEPS engineers intimately familiar with the miniaturized circuitry of avalanche beacons. The knowledge for constructing such electronic platforms could be adapted, the engineers felt, to a battery pack and control module and integrated with the jet fan motor and balloon.

"Thankfully," Gompert recalls, "the PIEPS dudes were really smart." In the abstract, the project's ambitions and outrageously innovative design may have harkened back to Gompert's early flashes of inspiration. Just trying to describe the JetForce's activation and operation brought forth quizzical, doubtful looks: strap a contraption on your back, ski into a steep field of deep snow to trigger a potentially lethal avalanche, tug on a lanyard to send an electrical current to start a fan spinning at a blinding speed, and then hope a bag the size of a truck tire inner tube would inflate—and save your life. Still, Gompert remained upbeat.

If there was anyone on the team who had earned the right for optimism, it was Coppi. And the prospect of completing a prototype of the JetForce pack tentatively but ineluctably invited the question: "Who's going first?" But even Coppi, an adventurous, expert backcountry skier and rock climber, was reluctant to be thrown into an avalanche field about to be released with explosives. So the team determined they would first try a dummy test of the device. If

Coppi was eventually to be the human guinea pig, he had to be at the helm of the initial field test. Which didn't, as it turned out, completely absolve him of being in danger's roaring path.

Coppi, Gompert, and the avalanche pack team contacted the avalanche bombing ski patrollers at Snowbasin ski area overlooking Utah's Ogden Valley. Snowbasin had been the site of much of the 2002 Salt Lake Winter Olympics' ski races, including the men's and women's downhill. A big resort with extensive groomed ski runs, Snowbasin also has some wild side-country appropriate only for expert skiers. Much of the avalanche mitigation at the ski area takes place in that side-country where repeated winter storms deposit unstable snowpack.

The team's request of the avalanche mitigators was hardly a liability insurer's dream: Could they bring a dummy with an experimental, avalanche survival device on its back to the ski area's side-country, have them set off a large avalanche using bombs, and, as the dummy plummeted down the mountain, allow an employee to trigger the survival device and capture a video of the result? Oh, and the employee would be positioned above the fall line for him to safely stay out of the direct path of the avalanche but close enough to get video—so he would need to be strapped to a tree. Given the unpredictability of an avalanche, there was a remote possibility that he could end up catching collateral turbulence from its path.

Most of the time avalanche mitigators, like Titus Case of Alta ski area, are deeply serious about handling explosives and setting off tsunamis of snow in the interest of protecting human lives. But they allow themselves, in rare instances, to get in touch with their inner boyhood love of pyrotechnics, deafening explosions, and the glee of provoking fearsome displays of nature. Kicking the dragon awake, as it were.

This turned out to be just the level of eagerness with which the Snowbasin and Black Diamond teams assembled on a foggy mountainside one morning. Coppi, secured to the tree with a radio remote controller in his hand, watched in awe as the bombs exploded adjacent and below him. The dummy, affixed to the experimental backpack, went rag-dolling down the slope. A wall of snow and ice, rising like an ocean wave hitting a reef, hurtled past Coppi.

Coppi saw the dummy being sucked under the way a human would in such a severe slide. As it slid down the slope below him, he activated the control signal for balloon inflation. A big, red, synthetic-skinned bag fought its way skyward from within the turbulent snow. In seconds, it was floating like a marker in a ship channel. Then something sensational happened. The dummy, like a huge cork being sucked into the trough of a breaking groundswell, shrugged its way upward with the balloon and popped to the surface. All this was captured on a GoPro camera strapped to the dummy's head. The group now had proof of their concept.

It might have seemed they were on the verge of the JetForce pack's final refinement and swift, subsequent production. But being tough on themselves is a standard protocol at Black Diamond. Their company culture is fierce. Because they produce life-saving products, even when success seems at hand, they go back to the drawing table to ask more questions. Unbeknownst to Coppi, he was about to assume the tester mantle the dummy had worn on Snowbasin's slopes.

The final hurdle for the JetForce's design pertained to the timing of the skier's triggering of the fan and balloon, and how long the inflation should last. With the rocket fan motor, the designers had the balloon's full inflation down to under three seconds. It could not have been any faster given the girth of the balloon, and it was much faster than required. But a different problem needed solving. Avalanches produce two major life-threatening events for a skier. The first is the initial detonation of the slide and the skier realizing what's happening as he or she suddenly sinks into the boiling snow as cracks open like crevasses across the mountain. At that moment, the skier would ideally react by activating the survival pack, hearing the *whoosh* of inflation, and feel the cushioning effect that urges the skier skyward as the avalanche gathers momentum. But there's no guarantee the skier will remain completely on top of an avalanche.

No natural phenomenon as potentially strong, reckless, and unpredictable as an avalanche can be perfectly plotted in its behavior. The mountain slope, the depth of the snow fracture, the debris like dislodged rockfall, and the sheer acceleration—all are variables that have a lesser or greater impact on the threat to life. While the JetForce team had built a backpack based on the goal of keeping skiers and snowboarders on top of an avalanche, there was the possibility a victim could end up buried when the avalanche stopped, particularly if the rider slid into a gully known to avalanche experts as a "terrain trap." Could the inflated balloon then become a liability? Could it be forcing a skier's face into a chokehold with the snow? This was the final problem.

Gompert knew the time limit for surviving the initial slide's duration was likely forty-five seconds. "Anything after that, you're not surviving," he notes.

Assuming the skier was buried and the slide had lasted forty-five seconds or less, the skier would face a mechanical dilemma. The backpack and its balloon would have kept the skier from caroming down the mountain while completely buried and being pounded by rocks and trees. But as things came to a stop and a final wave, or perhaps a second avalanche washed over the skier, he or she would be buried, facing the crush of both snow and balloon.

Gompert, Coppi, and the design team concluded they needed a way to program the balloon to deflate. A switch or command to automatically turn off the fan wasn't enough—the balloon would stay largely inflated, seeping air but still occupying too much space. What if, after a certain amount of time—say, three minutes from activation—the fan motor could be programmed to reverse itself at five hundred revolutions per second and suck the air from the balloon? Suddenly the skier would have two added effects enhancing survivability: a much larger air pocket around their head, and, with the space evacuated by the balloon, the ability to move their head and arms. It would be up to an avalanche beacon to guide rescuers to the skier, but the skier would be able to more freely move hands to clear away snow and to shout from under the snow pit.

The electrical engineers set about programming the backpack's circuitry to instruct the rocket fan motor to behave as desired. Now they needed someone to test it and experience the actual effects of being buried when the fan motor reversed, deflated the balloon, and left the victim to breathe and move under the snow. Coppi raised his hand.

THE TEST TOOK PLACE OUTDOORS. A Plexiglas panel, roughly six feet long, was installed like a transparent wall in the snow. Coppi, wearing a Black Diamond Avalung for safety—a kind of snorkel that sucks air from loose snow—nestled his body alongside the panel. His colleagues shoveled snow onto him until he was almost completely buried. Then Coppi fired the JetForce bag, heard the fan motor make its powerful *whoosh,* and felt his head become surrounded by the protective bag. The team furiously dumped more snow on him until he was completely submerged.

In a video that was shot with a camera up against the Plexiglas, all one can see are traces of the orange reflective lenses of Coppi's ski goggles. Then, at the three-minute mark, the fan snatches all the air from the bag, and Coppi breathes more easily and begins to move his head and hands. Celebrating the success of their unique design, the JetForce team was convinced their ingenuity had conquered any remaining reservations about their invention. But then Gompert's luck hit a speed bump.

BLACK DIAMOND IS AN INTERNATIONAL company. Among rock climbers, skiers, snowboarders, and mountaineers, the brand is revered. Known for expensive products, the company's quality standards are without parallel. But because it sells gear widely in Europe, all products must be certified by a process and oversight board that issues—or denies—the symbol CE, an abbreviation of the French phrase *Conformité Européenne.*

The CE marking means a product has been certified to meet strict European guidelines that conform with human health and safety standards as well as environmental protection. For a well-established company like Black Diamond, one might expect a simple, efficient road to approval. But the CE certification board puzzled over the avalanche device's fan-based theory and design. No one had ever produced an avalanche survival pack using a fan and complex electrical components programmed to work in this way.

The CE board threw challenges at the JetForce pack that frustrated and sometimes mystified Gompert and Coppi. The final straw arrived when the board dictated that there be a test of the backpack and its balloon inflation to prove it could create enough pressure to buoy the weight of a person—in water. Gompert and Coppi were now faced with devising the test as if switching from envisioning a skier on snow to a deep-water diver who wanted to jet to the sea's surface using the balloon and float on it. It seemed mind-boggling. But the board insisted they wanted to see proof the electrics would work underwater while floating a person. That was when the JetForce team looked at Coppi.

After finding a tank used for testing submerged equipment—essentially a huge dunk tank—Gompert and Coppi had it filled with four hundred gallons of water. Wearing the JetForce electrical harness and mindful he would be submersing it in water, Coppi dutifully went into the tank like a scuba diver testing a new aqualung. Pulling the activation wand, he felt the fan force water into a huge bladder on top of which he rode. This proved the fan could generate enough force for buoying a human above snow. And Coppi had not been electrocuted. The test pilot had notched another milestone.

"The electrocution risk was real," Gompert recalls, although safety protocols had been in place. Ultimately, the CE board did not compel Black Diamond to use the test as part of their certification, but "it was entertaining," Gompert says. If nothing else, the electronics had proven their short-term resistance to an invasive influence—water. "It's safe to say," Coppi declares, reflecting on the prototyping obstacle course, "I've deployed more JetForce packs than anyone."

The CE board at last awarded the avalanche backpack the necessary certification, and Black Diamond accelerated the project toward production. The team made final tweaks to the pack that might have seemed incidental to someone not intimately familiar with backcountry skiing. Instead of the usual waist belt and sternum strap one finds on most mountain backpacks, the designers added a crotch strap so the JetForce would not be easily torn off a skier's back by the extreme forces of an avalanche. In a practical touch, the crotch strap was colored red so a ski guide, out in the backcountry with clients wearing JetForce packs, could see from a distance if the strap was snapped securely.

The first JetForce packs were available to consumers in 2014. Black Diamond's website sold out the full stock in the first twenty-four hours. Early customers were mostly professional ski mountaineering guides and hardcore aficionados. At about one thousand dollars per pack, a Halo 28 is something of an engineering miracle, so price wasn't a barrier as it might have been with another product launch. If you have to use the JetForce at all, and you survive, the money is a throwaway proposition.

I listened to Gompert recount the saga of the JetForce pack—from the early days of sparklers in the mind to the sluggish days of blind-eyed bureaucracy. In some ways the Black Diamond team had developed more than an avalanche rescue device. They had confronted head on one of the main causes of avalanche deaths—asphyxiation under the snow. As engineers seeking the solution to a problem, they had homed in on identifying the essential need: they had built a glorious machine to invent air.

AT THE BLACK DIAMOND HEADQUARTERS, Gompert and Coppi walked me into a cavernous room that had been converted into the employee gym. It was a rock climber's get-in-shape Shangri-la. A huge climbing wall took over a quarter of the space, and other machines filled the gym—all geared toward building strength for biceps, forearms, and fingers for gripping sheer, granite walls. Muscular pain provided as a company benefit.

Either from forgetfulness or the sense of reassurance the Halo 28 JetForce pack provided, I still had it on. Gompert and Coppi urged me to give it a spin.

I held the button down to arm it, and the Buck Rogers–whimsical *whoosh* surged behind my head. I tightened my grip on the aluminum wand to my left, and Gompert and Coppi told me to "pull it hard." In seconds a huge red airbag haloed my head. I felt it but couldn't see it. The sound of the fan jet motor loudly thrummed away, sturdying the inflated life ring, fan blades spinning wildly at more than a thousand revolutions per second.

Within the prescribed three minutes, the fan reversed and the balloon swiftly deflated. I stood there like I'd just been ejected from a superhero's escape hatch. Before me sat Gompert, smiling, a little devilishly, like a clever crusader who'd beaten a foe. Because, after all, his favorite T-shirt is Abe Lincoln masquerading as Batman.

It's understandable why Gompert and Coppi are proud of their work on the JetForce. But they know the dynamics of the backcountry well enough to see there could be unintended consequences for some people who don the device. Their concern is that a small number of skiers might wear the backpack like an "invisibility cloak." In other words, the wearer might take more chances in avalanche-prone terrain on steep slopes because he or she has a false sense of being impervious to avalanches. Facing a pitch that's raised danger signs, a skier wearing a JetForce, says Coppi, might disregard the warning signs and think: "'Maybe I should ski it.' And that's the wrong mindset." But the lion's share of backcountry skiers, they feel, see the pack as another tool—like a beacon or shovel—to rely on when they have been prudent, done everything right, and still the white dragon comes to call.

Gompert shook my hand and headed off. Outside, Coppi and I walked toward my car. A steady, cold rain had begun. In the distance, impossibly high to be so close to Salt Lake City, was a mountain peak. It was flanked with trees and a thousand ribbons of snow painting its cliff bands. Up there, the icy rain was snow. Hovering above the gray clouds, the peak looked savage, foreboding, a place no prudent person would want to go.

Before I'd met Coppi I had been sent a photo of him skiing a couloir with a full Black Diamond pack, a JetForce. By the angle of his body, it was clear he was going fast and turning hard into each steep arc to keep his skis on edge. Snow-flecked rocks on either side of him were the height of cargo ships' hulls, menacing, looming over Coppi as if towers for witnessing his descent, gauging his worthiness to even be there. I mentioned the photo.

He pointed up beyond the parking lot to the pinnacle. “That’s Mount Olympus,” he said. He and his companions had skied one of the Memorial Chutes, a 45-degree slant. I looked back at Coppi, and even on such a gloomy afternoon, his face was aglow with memories of the ski run.

“We skied it early last Tuesday morning,” he said. “And I was at my desk by 9:30.

15

GEARHEADS: MAGIC BOARDS

Stephan Drake's skiing story started a lot like that of avalanche forecaster Craig Gordon, who began his Utah sojourn sweeping floors at an Alta restaurant. Like that of Titus Case, who, before spending forty-two years as an Alta ski patroller and avalanche director, worked at the Rustler's Lodge. Or Doug Chabot, director of the Bozeman, Montana, avalanche center, who waited tables, just after college, at Whiskey Jack's at Big Sky.

Drake's winter devotions began with him desultorily washing dishes at night in Las Lenas, Argentina. It may have left him with chapped hands, but the job gave him all day to do nothing but ski. Menial jobs launched him on an eight-year orbit, earning summer money in Colorado so he could jet south again to the Andes and ski lines like Eduardo's, a near four-thousand-foot run. This, Drake says, revealed his true future would begin with hunting down discarded skis in the "dust bins" of obscure ski shops.

Like the avalanche professionals, Drake was just a tenderfoot then, logging time in his ski boots, skiing day after day until his arches ached and his big toenails turned black. But this is getting ahead of things.

Before Argentina, there was Drake's boyhood in New York City, not exactly known for ski vertical; but even in New York, a boy could dream. "At an early age, I had a subscription to *Powder* magazine. I begged my mom for it." And there was his grandfather in Colorado Springs, a retired army colonel who had fortuitously bought a condo in Aspen. Every spring break, he had Drake out for two weeks to ski Aspen Highlands. There Drake became the de facto protégé of ski instructor Bill Gant, a "super classic ski bum" as Drake describes him.

Drake was smitten with Gant's old pickup truck, his dog that hung out in the truck bed, and with Gant's bushy, retro-hippie mustache. Gant ushered Drake to "secret powder stashes." Afterward, Drake's ski bum apprenticeship continued off the slopes. He "told me I should check out Pink Floyd's *Dark Side of the Moon*. . . . He instilled in me the aesthetics of skiing. . . . I could see the spiritual part of that, and the art of riding mountains."

But those dust bin skis awaited Drake. Living in Argentina with professional snowboarders, he began to glimpse his future.

Described by a ski resort guide as "one of the best in-bounds runs on Earth," Eduardo's sits at the top of Las Lenas, heli-ski–style terrain accessed by a rickety double chair lift. One day, Drake and his snowboarder roommate were perched at the top. Drake, riding a pair of skinny skis, dropped in first. Turn after turn after turn took him to the bottom, where he "collapsed in a pile of sweaty exhaustion." Moments later, Drake watched his partner come down Eduardo's, throwing huge rooster tail turns at sixty miles an hour. When he was within a few yards of Drake, he launched a big air, sailed over Drake's head, and landed the jump. That was Drake's moment of paradigm shift, his own private, deep, wide-eyed gawk into the snow globe of what might be.

"What kind of tools," Drake thought, "will let me ski that run in the way my mind imagined it could be skied?" Seeking the answer to that question sent him on a resurrection quest for old, early 1990s skis that had been designed for limited, even rare use. Atomic Powder Plus skis and Rossignol Axioms had been made almost exclusively for helicopter and snowcat skiing, particularly that practiced by Canadian Mountain Heliski, the pinnacle of such skiing in its day.

If you squinted enough, Drake thought, you could see those skis as snowboard influenced. But the design, he felt, wasn't fully executed for an all-mountain ski that would also excel in powder. He thought of snowboards and then about surfboards, which had always fascinated him. "I was committed to skiing," he recalls, "but I wanted the freedom of surfing and snowboarding."

As Drake remembers his early pursuits of vintage powder skis, he takes long pauses between his thoughts, more like a philosopher king than a skier bro. Another image surfaces, that of a musicologist fingering through old vinyl in a record store, looking for rare Velvet Underground, holding up black discs in the midday sunlight to study the scratches. Only in Drake's case, his eyes were canvassing ski sidecuts, his palms tipping the boards to feel the rocker, his fingertips rubbing the curves of ski tip shovels and the trailing design of tails. His quarry had been the outlaw skis and gear of an earlier time, designs like Volkl Snowrangers, or K2 Explorers, Italian-built Silvretta bindings that looked like silver jewelry, Voile free-heel skis that broadcast to other skiers "only hardcore here."

Starting with some old Atomics and Rossignols, Drake laid them on tables and began bending "rocker" into them, the kind of shapes he'd seen in surfboard profiles. Contorting the skis into a profile he liked, he'd stand back and look at the topsheet graphics. His quest for ski purity was overshadowed by "European race-oriented ski brands." He was looking for the rhythms of alpine nirvana and what he saw was meant for the show-off denizens of the Autobahn. So Drake started to paint the topsheets. Single, uniform colors, primary hues, and organic mountain tones. No buzzy illustrations like static-electricity charges, high-voltage warning icons, or hallucinatory dreams. Just absolute tranquility, in tune with his vision of skiing a perfect run down Eduardo's.

What Drake lacked, though, was scale. He could revive a few pairs of skis for himself and some friends, and the result felt like progress. But to explore the ski aesthetic he'd imagined for a run like Eduardo's—one he'd first sensed as a kid when discovering Aspen powder stashes—he would have to make a bigger leap.

It was then that Drake, a onetime dishwasher who'd become a fugitive ski designer, let himself freefall into his future. He decided to start a ski company.

SEVENTEEN YEARS AFTER GETTING INTO the ski design and fabrication business, Drake is the founder and leader of Salt Lake City–based DPS Skis (short for DrakePowderworkS), Sitting in a meeting room at DPS, he seems unchanged by his corporate role of nearly two decades. This, despite the fact that on the wall are three clocks labeled SALT LAKE, STOCKHOLM, TOKYO, time-telling harbingers of approaching deadlines. Relaxed and pensive, he's still the snow drifter in a green-plaid flannel shirt, and trousers that look almost stone-washed but from genuine age.

Part of Drake's inspiration for starting a ski company emerged from not being able to find the kind of skis he wanted for the mountain lifestyle he was living. That included a global tribe of friends pursuing his own aesthetic ideals. He calls them "the tip of the spear" of the sport, big mountain chargers in places like Las Lenas and Alaska's Chilkat Range. "We were after the perfect turn, on the perfect run," he recalls.

In a brochure for DPS's 2018–19 winter season, Drake wrote a short essay, "The Perpetual Dream," to capture this lifestyle. "The small room where you live is cozy. Wood smoke is perfume. Outside, the days are full of clean and crisp alpine air. Pushing your craft, and surfing through the deepness, is the way. . . . Snow piles deeper and dryer, and nights pass easily without much concern for the next day. There are friends and laughter."

Creating ski equipment that blended with this lifestyle wasn't easy. What was accessible were skis designed for the traditional style of powder skiing, a technique Drake describes as "a plow and sink job," a style of skiing that exacts more physical exertion from the skier as edges are dug into the turns and snow. He sought skis that, softly cajoled, could naturally float through turns and snow. He imagined skis that carried the skier without overtaxing calves and thighs, skis that though lighter weight could still build a lot of speed to reduce the force a skier invested in each turn. He wanted less bounce and more nuance in the turns. Drake's way of thinking about skis derived more from hydrodynamics than the brute physics of gravity.

Recalling his snowboarder roommates, Drake invokes the "f" word that drove a lot of his inspiration—freedom. "I was incredibly envious of the freedom they had from the fall line." In 2001, he partnered with Cyrille Boinay, a former member of the Swiss national ski team. From the beginning, the foundation of Drake's designs was carbon fiber. He was convinced it was one of the keys to the lighter weight he was after. He had seen the material infuse advances in many other sports—with tennis rackets and golf clubs. Drake wondered, why not skis?

But getting from concept to reality proved daunting. Drake first launched his production operation in China. He ended up having to live there six to eight months a year for five years, and his budget was the proverbial shoestring. "It took a decade to figure it out." During that time, his partner decided to leave the company and Drake continued on solo. Design and construction innovations progressed, and Drake finally realized he needed to bring everything in-house in Salt Lake.

"Once you get into this thing there is a constant dissatisfaction," he says of design. As DPS adjusted rocker, width, and the shapes of tips and tails, he recalls, "I knew we were onto something." Making turns on prototype skis, the G-forces would be powerful, but the physical effort was balanced. He glimpsed that nirvana, "the powder moment." Giving the skier more of those transcendent ski moments is Drake's ultimate pursuit for his customers—and himself. To call him a "purist" in his bid for ski perfection is almost to cheapen the description; it's a two-dimensional word when Drake works in at least a three-dimensional design mindset. It's like describing someone as "religious" when they are pursuing Sufi mysticism. He's actually striving for a fourth design dimension, something that happens when a ski, a rider, and the snow combine in an ineffable synthesis.

"I can remember runs—in my head I have my top five or ten runs. . . . And when you come out the other end, there's this crazy euphoria. . . . Skis are a tool that interact with you emotionally." In his brochure essay, Drake seeks to define that emotional bond: "Every run, every turn expands the soul via the seamless transfer of power and feeling. Equipment reflects the art, and it's all so good that your movements become universally dynamic and economical. Your body doesn't hurt—it's perfectly in sync."

WHEN DRAKE PONDERS NEW SKI designs, he thinks about the backcountry, where he spent the lion's share of his time traveling with his global ski tribe. Now he strives to serve both those backcountry touring zealots and the resort skiers who have a similar affinity for chasing powder stashes. He estimates that "at least fifty percent of our customers" have some involvement in backcountry skiing. And he expects that to increase. "It's certainly growing and the equipment—ski boots, bindings, skis—the equipment has facilitated that. . . . When I was growing up, the gear was hard to use, flimsy. A lot of those barriers have come down."

DPS's ski offerings reflect the balance in skier styles that Drake seeks to match. Skis fall into three design and construction categories—Alchemist, Foundation, and Tour1. Alchemist skis are heavily carbon-fiber based, mostly pure powder and big mountain boards; Foundation models are carbon/bamboo/fiberglass, built as all-mountain skis ideal for resort skiing; and Tour1 skis are especially lightweight for backcountry touring and uphill travel on climbing skins.

While Drake's own skiing in places like Alaska's Chilkat Range required an uncommon level of backcountry exploring, he believes time on resort runs has its definite place. "The resort is the place you get your repetitions in. It's harder to do that in the backcountry." His experience in big mountains left him with insights he brings to design refinements, even down to the level of snow sluff management on high-angle terrain and the ever present threat of avalanches. Like most big mountain pros, Drake has seen friends die in the backcountry.

One friend and colleague, Rob Lieberman, a resident of Telluride, Colorado, worked as a guide in Alaska. In March 2012, Lieberman, then thirty-five, guided a group of six skiers and snowboarders in an area called Takhin Ridge, north of Haines, Alaska. Five feet of new snow had fallen and an overnight windstorm loaded up peaks and bowls with loose, huge pillows of snow. An avalanche broke, trapping Lieberman and a client. Both died. "I never met anyone more passionate about powder skiing," Drake recalls.

In some ways, Drake walks the fine line of a powder-skiing impresario who knows there is a hungry audience for his ski designs—those seeking the outer limits of powder plank performance—but who must simultaneously acknowledge that some people craving his skis, like owners at the wheels of exotic sports cars, are going to ignore things that can get them killed.

Nick Paumgarten, a longtime staff writer for *The New Yorker*, has written widely about skiing, calling some backcountry lapses in judgment attributable to the "fog of pow," an overwhelming desire to ski fresh powder snow even when hazards are clearly evident. In a brief, December 2012 piece, following Washington state's Tunnel Creek avalanche deaths of three skiers at Steven's Pass, Paumgarten noted that ignoring the "array of immeasurable risks" in backcountry skiing almost certainly causes even the most experienced skiers to sometimes choose the imminent elation of powder turns over a prudent—but tiresome—hike out of danger's way.

But Drake and his cohorts are in the business of building tools for people whom they hope can weave fierce ski lines while still drawing on deliberate expertise, despite the tension between the two. And yet, Drake clearly prefers inspiration over trepidation. DPS even has a film arm called DPS Cinematic. One of their recent releases about Alaska heli-skiing is titled *Predators of the Northern Deep*. And the company's 2018–19 manifesto isn't ambiguous: "DPS is about the mystery encountered during a slide across deep snow. It fuels a reverence for mountains, storms, and the people who sculpt powder culture. It leads us on a search for distant ranges and on a quest for the most progressively shaped and built skis on the planet."

DOWN A HALLWAY AT DPS headquarters, and into a workshop filled with machines and mostly young men, the evolution of carbon-fiber skis is taking place. An amalgam of thick, fabric-like weavings are being suffused with a black goo, like extruded licorice. A single man stands in a small anteroom saturating the long, skinny black swatches with widths mimicking ski profiles. It is a job, Drake says, "no one really wants." But the man smiles behind his respirator and waves.

Next are short stacks of Aspen wood ski cutouts, trimmed to the lengths and widths for DPS's range of ski categories and skier genders. Drake directs me to another pile of cutouts from balsa wood, a lightweight core for backcountry touring skis that make them easier to lift and power forward in deep snow.

A finished ski that is a sandwich of wood core, trimmed carbon fiber, steel edges, and base material sits in a jig awaiting a pastel-colored topsheet. The layered ski has clamps around its perimeter to help adhesive set up tightly. One might imagine an automated machine, rather than human hands, to achieve this. "People think the manufacture of skis is really automated, but there's a lot of handcrafting," Drake explains. The sistering of all these materials, depending on the intended use of the ski, has a definite effect—certain weights and densities of wood to dampen the ski so it won't get "anxious" on bumpy snow or ice. Variable after variable leads to a final equation that Drake seeks for riders. That area of the production facility looks more like a New England boat-building barn than a high-tech ski atelier at the foot of the Wasatch range.

Evident in all the designs, and the production genetics they share, are Drake's signature refinements—the surfboard-inspired rocker, the spoon-like tips, minutiae of sidecut. As he touches the skis like talismans, he talks about DPS's progression over the years. "The boundaries of shaping have been touched on," he says. "I'm excited about the last ten to fifteen percent." The company's current focus will "be less on radical leaps and more on perfection of what's been proven." Lately, Drake has been working on final tuning of skis, making sure the steel edges are shaped and ground perfectly to work with the near frictionless glide of base layers. "Our pursuit of perfection has made us focus on tuning. You can make a great design with a bad tune and it will ski horribly."

Just beyond the tuning area is a small kingdom of finished skis, filling shelves, awaiting shipping, locked inside chain-link barriers. The whole place is a veritable Willy Wonka factory of candy-colored skis, exquisite as rare works of sculpture. At a final inspection station, a man hovers over a pair of skis while whistling to a Neil Young song. The soundtrack doesn't seem an accident. A kind of 1960s soul, updated with the revolutionary properties of carbon fiber, infuses the DPS building. It's as if the early California spirit about surfboards handmade in garages—and the spirit of Drake's Aspen Highlands "super classic ski bum"—have been transported to twenty-first century Salt Lake.

By the entrance to DPS, Drake pauses beneath a huge photograph, maybe four by six feet. It's a helicopter vantage point, shot from high over a mountain, a panoramic view toward a slope that's like a blown-up labyrinth of snow and blue ice. A single skier makes a sweeping Daytona turn to skirt a crevasse and connect with a snow spine as ghostly as the neck of a giant, white dragon.

The skier in the photo is Drake in the Chilkat Range. "It's wild up there," he says eyeing the image. "We ended up exploring it with snowmobiles." For a moment he drifts back there, a far cry from the DPS factory floor. As he says in the brochure essay: "Our factory is located off a diesel-trafficked industrial road, near a body shop, beside train tracks. A sole east-facing window frames the Wasatch, as the mountains rise dramatically from the valley floor. Inside, we toil daily to constantly improve; to push and push, and push some more."

On the wall above him, Drake has a window into another world, a piece of the same one he still has inside him: that of a young ski vagabond, washing dishes and dreaming of Eduardo's.

16

SOLDIERS OF THE SNOW: GUIDE VOICES, HASTY PARTIES, AND RESCUE DOGS

Backcountry ski guide Shannon Finch remembers watching a ski client get swept away in an avalanche on a mountainside in Thousand Peaks Ranch, where she works full-time for Park City Powder Cats. One second, the client was standing beside her. The next, he dropped into a run, made a turn or two, fell, and slid ten feet down from one of his skis that had released.

Another client, a snowboarder, became impatient while the skier was trying to retrieve his ski. "Hey, I want to go," the snowboarder told Finch. As he adjusted his snowboard binding, he unintentionally slipped into the top of the pitch, slid down to a shallow rock, and started a small avalanche.

Finch glanced down to where the fallen skier was struggling. Below the breaking avalanche, he was swiftly caught in the onrushing snow and heaved over a hump in the slope. Then he vanished. "The last thing I saw was the client's arm go up and then disappear," she recalls.

Hearing Finch describe the experience, it's as if she had set something down on a beach berm only to see a giant wave suddenly snag it and suck it out to sea. But this was a human, who had just paid five hundred dollars for a day in which Finch was charged with shepherding him safely through mountain terrain. She ordered the client group to freeze in place. She jumped on her skis down into the avalanche debris field and skied over the roller in the slope. Just below, the client came into her sight. He was mostly above the snow, digging himself out.

"I felt a surge of emotion," Finch says. "I just remember asking, 'Can you breathe, are you okay?'" The "vibe" of the client group changed dramatically that day to a wary regard for the potential dangers around them, although all agreed they should keep going. After another five or six runs, Finch and her fellow guides, back at the Park City Powder Cats cabin, debriefed with the client. "You look at these things," Finch says, "and ask, 'How did we get here?'"

The next day, she had ski patrol duty at Sundance ski resort. But after a night of replaying the avalanche accident in her mind, the image of it haunted her. She couldn't shake the feeling of "What if?" Suppose her client had been completely buried and seriously hurt—or worse. "I went to Sundance and just felt hollow," she recalls.

Unable to stop thinking about the what-if, she had a long talk with her twin brother. And then she started to cry. "I needed to let that emotion out. I had had all that adrenaline." Finch, who has had mercifully few experiences with avalanches, found her scare to be an emotional jolt that she learned from. Like other snow professionals—particularly patrollers and guides to whom skiers and boarders entrust their lives—she discovered the true depths of her own commitment and weight of that responsibility to be even greater than she imagined.

During the ten years Finch has worked as a patroller at Sundance, and more recently as a backcountry guide, she has developed a skill-set beyond being merely a bearer of a first-aid kit and an assessor of avalanche risk. She has developed what she calls her "guide voice." One lesson learned did not come while in the mountains but while guiding kayakers, which she does in summer. "I remember specifically a time when I was working with a Kiwi guide; we were providing instruction to a group of our kayaking clients," Finch recalls. "During a roll session he accused me of just being a 'cheerleader' to our guests. I was put off by the comment, but it allowed me to reflect on my guiding techniques." As she thought about it, she realized she was sometimes voicing a message of encouragement, when it really needed to be an emphatic warning about a potentially dangerous outcome.

"It is important as a guide to have the ability to use your voice in a respectful but commanding way; not everyone will be your friend at the end of the day. You may not make the most 'popular' decisions when it comes to meeting client expectations, but safety and good risk management will always trump popularity for me." Her "guide voice" has become an instrumental tool for her work in the backcountry. "My guide voice is clear and commanding, and I have found through experience that this voice has grown stronger over the years, and that's simply come through experience, close calls, near misses, and trying to manage groups. Managing and guiding clients, both on the water and on the snow, is a combination of understanding group dynamics, expectations, and ability."

A Utah native, Finch grew up in Orem, one of five siblings. "We didn't have a lot of money," she says of her childhood, "but skiing was really important in our lives." In her mid-teens she competed in freestyle skiing competitions. Just before turning sixteen, she experienced her first exposure to the beauty of the backcountry and was hooked. After studying recreational management at Utah Valley University, she took a job at Sundance "making coffee" and doing administrative work so she could get a free season pass. By the time she was twenty-one, Finch had completed her EMT certification and earned an invitation onto the ski patrol. She began a focused process of learning about avalanche mitigation and mountaincraft. After nearly a decade of resort patrolling, in the winter of 2015–16 she began to guide in the backcountry.

Guides like Finch comprise an elite cadre of snow professionals. Whether guiding clients on climbing skins, in snowcats, or on snowmobiles and in helicopters, backcountry guides have to bring to the mountains the rescue and medical skills of resort ski patrollers and then match those with the highest level of avalanche knowledge and terrain familiarity. There is the ever present task of beginning a day with ten or twelve clients, some of whom may be strangers to each other, and having to read personalities quickly to judge where conflicts may arise that heighten risks. Overconfidence, or a lack of confidence, can create short circuits in communication. When guides like Finch see directives and signals breaking down among a group, they know the accident threat level is spiking.

Finch, who looks younger than her thirty-one years, wears her blond hair in a braid that spills out from under her knit cap. She sometimes has to deal with age and gender dynamics, which can make her duties even more complex. "Typically, in the snow professional industry, I haven't felt it as much," she says. "But I have felt it from clients. They expect me to prove myself." There are older men, for example, who insist on removing their skis from the basket affixed to the side of the snowcat as if to imply she isn't strong enough. There are overanxious skiers and boarders determined to conquer the highest peaks, oblivious to avalanche threats, when Finch feels such a decision would be dangerous for the group as a whole. In these cases she's felt the suggestion that being female and young—an assumption she's more timid and less experienced—is preventing her from letting a client "go big."

In developing her "guide voice" and her ability to firmly communicate instructions, Finch has honed a quality amateur skiers may well lack when faced with group pressure, or groupthink, which is what one Canadian guide calls a lack of the "social courage to speak up." Guiding and patrolling have historically been male-dominated roles, so someone like Finch has to work a little harder to establish her brand as a guide. "We all have a little ego," she says. "It exists in our culture. Wanting to grow together makes it easier to work and grow."

Unlike young skiers and boarders who may go into the backcountry without gear and training, groups like those who ski and board with Park City Powder Cats are coached through a day in the out-of-bounds. Avalanche beacons are issued in the morning. One of the two guides aboard the snowcat is designated "lead" to test a slope before clients go and the other skis "tail" in case anyone falls and needs help. Finch, like many snow professionals, has witnessed a significant increase in the number of people who want to veer from resort skiing into the out-of-bounds. "It seems like it's trending right now," she says. "I certainly see it in the social media world. The trailheads are full."

ONE DAY, EARLY ON IN her work at Sundance, Finch saw a ski patroller streak by, followed by a dog porpoising up and down through the snow. The dog wore a red ski patrol vest. "I thought, 'I want to do that,'" she recalls. "I want to be like her." After joining the ski patrol, she couldn't stop thinking about the avalanche search dog she'd seen. She noted several patrollers had them. "I was in the right place at the right time," Finch recalls, "because they needed another dog."

Ski patrollers called to rescue someone caught in an avalanche are acutely aware of the math that works against them once a person has been buried. That's why they call such rescue responses "hasty parties." In many ski resorts, and on backcountry rescue operations, like Utah's Wasatch Backcountry Rescue, search dogs have become critical members of the team. In 2011, Finch began looking for a candidate puppy. Finch sums up the criteria search dog handlers use when describing a good candidate: "It's smart and it wants a job." Oversimplified perhaps, but it makes its point. Working dog breeds, like those bred for farms or retrieving birds on hunts, often fill search dog roles. Finch learned about a breeder in Nebraska who had just overseen the birth of a litter of twelve English shepherds.

Working by phone and email, Finch monitored a series of evaluations, conducted by the breeder, that had been developed by a couple named Wendy and

Joachim Volhard. Eleven characteristics are graded in the Volhard tests involving qualities like a dog's "social attraction," how well the puppy connects to people; "following," a willingness to exhibit loyalty to a person; and "retrieving," how willing the dog is to do something for a person. After putting the twelve puppies through videotaped sessions of Volhard tests, the breeder sent Finch the videos. She and the breeder reviewed them, then selected four puppies as final candidates. From those, she chose an eight-week-old she named Leif, and her relationship with him and his training began immediately.

Avalanche search dogs go through an evaluation and training process that's nearly as rigorous as the one for ski patrollers. Over time, the handlers have a learning curve along with the dogs. Initially, the dogs are just candidates, exhibitors of qualities that suggest a certain aptitude but unproven in real mountain conditions. At ten weeks, Leif was put on ski chairlifts, rode on snowmobiles, and was taken up for a flight in a helicopter. By the end of the ski season, search dog trainers working with Finch had seen Leif show signs of loving loud motors, deep snow, and stiff winds in his face, so he graduated to the next level. That meant Finch and Leif would begin, the following winter, on a demanding pathway all search dogs and their handlers travel. It's one a small number of snow professionals engage in because of its expense—one handler estimates about twenty thousand dollars—and the potential for failure.

FORMER MARINE OFFICER JOE JENNINGS retired in 2004 with his wife, Betsy, to Utah's Ogden Valley, mostly because of the area's world-class skiing. There he worked as a mountain guide at a valley resort, Powder Mountain. A few years later, wanting to get a dog for walks in the mountains, Jennings started to look. "All we wanted was a good, healthy, well-socialized dog," he recalls. "We found this breeder up in Wyoming who had a good reputation." Jennings bought a golden retriever he named Gunny after a certain breed of Marine Corps sergeants. But what he didn't know, until soon after, was that the breeder had been raising dogs for hunting. "This puppy was driving us nuts," Jennings says. "If he got bored, he would drive us really crazy. And he got bored really easily."

Telling a Powder Mountain colleague about what he was experiencing with the dog, Jennings's friend said, "Maybe he needs a job. Why don't you bring him up here and let's test him for being an avalanche dog." On a Saturday morning, with the temperature at 10 degrees Fahrenheit, and a 20-knot wind blowing at the mountaintop, Gunny jumped from Jennings's car "and just dived into a snowdrift." The friend, who had trained other avalanche dogs, ran Gunny through the same tests Leif had gone through. At the end of the day, he told Jennings of Gunny's suitability for finding people buried in avalanches: "That's what this dog was born to do."

Jennings and Gunny began training for the first year's K9 certification. At the time, they were affiliated with a group called Great Basin K9 Search and Rescue (SAR), and there was a single but complex test Gunny had to train for. After many weeks of rigorous training in the snow, Jennings and Gunny prepared to undergo the evaluation. "I probably had more to learn than he did," Jennings recalls.

Avalanche dog training derives from a model developed in the Swiss Alps where dogs have been used for generations to search for avalanche victims. The training Finch and Leif had to undergo was a certification overseen by Wasatch Backcountry Rescue (WBR), a different search organization than Jennings's. Over two seasons, WBR requires the dog to surmount four levels of difficulty for performing searches tantamount to what the dog would have to do in a real burial situation. For Leif, the training began by raising his awareness of snow mounds—simulated avalanche debris that contained a toy Leif wanted. Then the simulation advanced to holes dug in the slope and eventually to small snow caves. "As his handler," Finch says, "I take his favorite toy and I climb in the cave and he comes to me. Then we make it harder for him to get in." Each time Leif succeeded, he received a reward. Once a dog masters the simpler, more congenial levels of finding someone in the snow, the training begins to involve strangers in snow caves. From there, it evolves into finding pieces of buried clothing.

The first year for Leif and Gunny culminated with a search test on a field of snow one hundred yards square, where three people had been partially buried, and several patrollers were probing the snow while skis and backpacks were scattered around as potential distractions to the dog. For Leif and Gunny to move ahead, they had twenty minutes to find the buried people. "It took him twelve minutes to do it," Jennings proudly recalls of Gunny. In his Level B test, Leif had to find two buried people in no more than twenty minutes. He dug out the first one in seven minutes and fifty seconds, and uncovered the second simulated victim at nine minutes and seven seconds into the test.

After these initial tests, the dogs often graduate to a Level A test as used by Wasatch Backcountry Rescue and adopted by Great Basin K9. While this latter test involves much the same burial simulation as Level B, instead of people, who throw off a strong scent, three pieces of clothing are used. For Gunny and Leif that meant finding scraps of material such as wool sweaters, buried eighteen to twenty-four inches under the snow. The dogs had twenty minutes. "The intent is to simulate a deeper burial," says Jennings, noting that Gunny "did it in nine minutes." In his Level A test, Leif was tasked with finding two buried pieces of clothing; he uncovered the first in three minutes and seventeen seconds, and the second scrap at eleven minutes and forty seconds.

With Gunny certified for operations at Powder Mountain, and Leif for Sundance, the dogs donned their red K9 patrol/avalanche jackets and could regularly be seen riding the ski lifts. But training for avalanche search is neverending for the dogs as there are so many variables on ski mountains. As the simulated situations become more arduous, Finch says, some of the burden for success depends on how the handler executes. Positioning the dog to better receive smells on the wind, for example, is one tactic. Because weather can change by the minute, the dog has to become accustomed to fluctuating gusts, sudden blizzards, and distractions from machinery, peoples' panicked voices, and the frantic motions of searching the snow.

The success rate for an avalanche search dog to find a skier in a real burial can vary widely, as it does with a human searcher, depending on location and time. "If someone in the resort is covered, and the dog is at the top of the mountain," Finch says, "we have a very good chance of finding someone." In fact, she says, the mission statement of Wasatch Backcountry Rescue includes "a committed goal of the Live Find." But proximity of the dogs to an avalanche is critical. "The problem with dogs is getting them there quick enough," Jennings observes. "If an avalanche occurs in the backcountry, and your buddies can't get you out, the dog's going to be coming in for a body search." Jennings says Gunny has been in about seventy searches over the years, some in the snow, some in the backcountry in summer when hikers have gone missing.

"Dogs are like anything else," Jennings says. "They are another tool you can use. They aren't perfect, but they can do some amazing things." Finch has had Leif involved in searches when someone thought a missing companion might have ventured beyond the boundaries of Sundance into the path of an avalanche. Leif was taken to the spot but gave no indication of picking up a body scent. A second dog also failed to detect anyone. "It just comes down to knowing your dog and trusting him," she says. "I've learned to read that dog really well."

Now eleven years old, Gunny has retired from formal search duties. But Jennings says Gunny loved his time on the mountain in the snow and understood, as much as any patroller, what his job was. "He will hunt until he drops," Jennings says. "That's what he was born to do." Working with Gunny on ski patrol, even in miserable conditions, made Jennings's own job more satisfying. "He's much nicer," Jennings says, "than any gunnery sergeant I ever met."

17

MASTER OF THE OUT-OF-BOUNDS: ANDREW MCLEAN

Most world-class mountaineers reserve their highest level of wariness about safety for icy pitches on mountains that require the steel-tipped blades of crampons to punch, inch by inch, ever higher on ascents. So it was with an odd mix of quizzical stares and a momentary measure of wonderment that Andrew McLean and three companions, standing on a low-angle Tibetan glacier on October 5, 1999, saw something astonishing in the distance. Stunned by the sight, they could only stare, witnessing the beginning of a catastrophe.

There on the ice-strewn terrain of Shishapangma, one of the world's tallest mountains, was a chain reaction of collapsing snow and ice cornices that began about five thousand vertical feet above them. One after the other, huge cornices were breaking loose, as if the first glittering chandelier had snapped from its ceiling anchor and, freefalling, crashed domino-like into another crystal corona and then another, and another, until a gargantuan wave of frozen glass was sweeping onto the glacier, racing toward where the men stood in a state of disbelief.

Nearly paralyzed in the moments before impact, the four men ran. World-class mountaineer Alex Lowe dashed down the slope as if he could outrun the onrushing avalanche. A second expert climber, Conrad Anker, darted to his left, straining to reach the glacier's margin before the frenzy of snow hit. A third, cameraman David Bridges, darted down the snowfield like Lowe. McLean, far off to the left, angled down, tracing fifty feet farther left along the edge of the glacier. There he spotted a rock outcropping. Just before diving behind it, he glanced back to see a "massive powder cloud with [snow] tongues shooting out of it."

In mere seconds the catastrophe reached its full, terrifying furor. It was as if the roaring sky had descended onto the glacier, turning day to night. "I knew it was a matter of seconds before I got hit," McLean wrote in a May 2016 recollection of the accident for the online editions of *Alpinist* and *Backcountry* magazines. "Burying my head and taking deep breaths, I tried to remain calm. Seconds later, the force of the air blast smashed my face into the rock hard enough to break my goggles and fill everything with snow. After a few seconds it dissipated, and I was ecstatic to find myself alive."

When he emerged from behind the rocks, McLean witnessed a sea of avalanche debris as he stared into a blindingly white horizon. The breath of the avalanche hung in midair, a settling exhalation from a spent cataclysmic force. Then he spotted a single figure. It was Conrad Anker, staggering at the margin of the decimated glacier field. When the two men reached each other, McLean could see Anker's bloodied face and eyes. "They're gone, they're gone," Anker could only tell him, McLean recalls.

So vast was the debris field and so deep and heavy was the snow, McLean says there wasn't "even a sense" of action to try to mount a rescue. The day's hike onto the glacier had only been exploratory; no skis or avalanche beacons had been packed. The crevasses that had been visible in the glacier were now filled with snow. Whether buried, or swept into one of the crevasses, the two missing men could not have survived.

Although they searched for their friends' bodies for several days after the avalanche, they found no trace of them. They returned home to the United States, abandoning their attempt to ski the 8,027-meter peak. Not until April 2016—more than sixteen years later—were the men's bodies accidentally discovered by two climbers. "Unfortunately, Alex and David are not the only two friends of mine who have died in the mountains over the last few decades," McLean wrote in his May 2016 essay. "I'm superstitious about removing any of my deceased friends from my phone or email list and, after a while, it has become a good reality check. Skiing and climbing can deliver the highest of highs, but it comes at a price: these sports are definitely dangerous."

While McLean never returned to the Himalayas, he has persisted on in his steep skiing exploits. "When I've pulled off a big mountaineering challenge," he wrote, "I feel as though I've experienced the stuff of dreams: life seems much more exciting and vibrant. I can fully relate whenever extreme athletes say, 'I don't have a death wish, I have a life wish.'"

MY INTRODUCTION TO ANDREW MCLEAN came as an email from Craig Gordon at the Utah Avalanche Center: "Andrew is a colleague, world-class mountaineer and guide, and a world-class gentleman." Gordon's words, I would soon see, proved faultless.

I met McLean at the Moose Café, a diner-style restaurant near Parley's Summit on the edge of Park City, Utah, where McLean lives with his wife and two daughters. When he entered the café, he didn't need to rely on his considerable mountaineer's route-finding skills to locate me. I was the only one in the place. While McLean and I had never met before, even if the café had been filled with people, I would have easily recognized him from magazine photos.

McLean ambled over to my table, we shook hands, and he slid into a chair with the same ease he shows whether hanging on a fifty-foot rope from a cliff or making a jump turn into a forty-five degree, seemingly vertical ski slope. At fifty-six, he is something of a dean in the ski mountaineering world. He had come to talk with me about the current state of his sport, about the backcountry, and about risk, consequences, and avalanches. A little bit of a "last man standing" among his generation of steep skiers, McLean has known ski partners who have died in the mountains in the Sierra, the Wasatch, the Himalayas, and other ranges.

Unlike some world-class mountaineers who throw imposing physical shadows, McLean is a compact man, easy in his skin, clearly confident but not inclined toward intimidation. His shock of dirty blonde hair and his youthful eyes give him the look of an aging prep-school boy who never got the memo about banking, law, or medicine. A onetime product designer at Black Diamond in Salt Lake, McLean now weaves together a work life of guiding ski mountaineering trips around the world, writing articles for magazines like *Backcountry* and *Alpinist,* and working on mountaineering product design.

"I think I'm a member of the gig economy," he says. McLean laughs easily, a quality that finds its expression in some of his writings, especially in *The Chuting Gallery,* his twenty-year-old guide to skiing eighty-eight classic Wasatch ski mountaineering chutes. With names like Homicide Chute, Suicide Chute, and Hellgate Couloir, the ski runs outlined in the book are the stuff of most skiers' nightmares. In a rating system he employs for each chute, McLean assigns such designations as: "Slopes between 45–50 degrees. You'd be lucky to live through a fall," or "Slopes continuously steeper than 55 degrees. Painful death from falling highly likely." In a nod to running out of ways to say, “You fall, you die,” he just aw-shucks-it and says, “Sixty-degree slopes. Just plain ol’ steep as hell.”

When McLean talks about his many narrow escapes in the backcountry, it’s as if he’s in a teaching moment, not seeking to stir awe or horror in a listener. “I think what doesn’t get mentioned a lot,” he says, “is how many times I’ve turned around.” He mentions one of his “projects” in particular, a Little Cottonwood Canyon climb and chute ski descent called the Hypodermic Needle, the name alone evoking visions of a spire of dread.

McLean says he failed to make the climb to the top on his first try, then reached the pinnacle on his second but found the snow too deep to ski. On his third bid, he made it. “You have a lot of failures,” he says, “but it’s still skiing and technique honing.” Almost all of McLean’s *Chuting Gallery* routes are in the backcountry, he notes, a defining quality that tends to be “self-eliminating” for a lot of skiers. And the routes often require multiple efforts. “It takes a lot of work,” he says of the process.

McLean has come to know risk so intimately that it’s another familiar tool in his backpack, like an ice axe forged from close calls and what he’s learned from them. Some risks he sees in the mountaineering world leave him shaking his head. Of the recent free-solo climb of Yosemite’s El Capitan by Alex Honnold, without using ropes or other safety gear, McLean says that just thinking about it “makes my palms sweat.” Of Honnold, whose historic climb was chronicled in the film *Free Solo*, he says, “He’s so familiar and comfortable with it. I think it applies to skiing too. If you’re familiar with it,” the risk seems manageable. “As far as what I’ve skied, I don’t think I’ve gotten close to the bleeding edge.” When other skiers skirt that edge, or even slip over it, it’s not lost on McLean. “When people push it too hard, I probably ski it, then will not ski with them again.”

Looking into McLean’s life and assessing the legacy of his ski descents, many of them covered in *The Chuting Gallery,* other skiers don’t always view the risks from his perspective. Noah Howell, a sometime ski partner of McLean’s, recounted in a spring 2018 *Backcountry* magazine article his first exposure to McLean’s guidebook. It was a slide presentation McLean gave at an REI store right after the book was published.

“Honestly, it freaked me out,” Howell recalls. “I was like, ‘What the fuck? Why would you want to ski this stuff?’ ” Then Howell proceeded to undertake skiing all eighty-eight chutes. In an outing akin to climbing and skiing with the professor emeritus, Howell and his brother made an attempt to ski with McLean the northeast couloir of a Utah vertical pitch named the Pfeifferhorn. On display was not just McLean's ski mastery but also his caution. In his book McLean describes the couloir's angle as fifty-two degrees, and "more fun than . . . sticking paperclips into electrical sockets." McLean went first, descending by rope. A few minutes later, Howell saw him climbing back up.

"'Umm, it's really steep and icy in the choke,'" Howell says McLean told them. "'I'm just not into it, but if you guys want to ski it, you can.'" Howell recalls, "My brother and I looked at each other and said, 'Hell no.' That's like Rambo running away from a gunfight."

In a section on avalanches at the beginning of his book, McLean notes: "This is a definite reality of big game chuting in the Wasatch, where the hazards of skiing steep slopes pales in comparison to the avalanche hazard. Unlike climbing or skiing where the hazards may be blatantly obvious, avalanche avoidance is a blend of art, science, experience, intuition and a bit of luck. Try to take as much of the luck element out of it as possible." When asked about the intuition element in avalanche detection, McLean ponders it for a moment. "So many times it's hard to say what you feel uncomfortable about," he says. "You start to develop a Spidey sense."

Born in the Pacific Northwest, McLean grew up mostly skiing a resort called Alpental, where he did some racing but eventually realized he wanted to blend his passions for skiing and rock climbing. A geographic detour in the mid-1980s for college at the Rhode Island School of Design took him away from the western United States. But once he'd completed his degree, McLean headed to Utah, where he eventually landed a job at Black Diamond. He spent thirteen years there before chasing steeps, early in the morning with his "dawn patrol" gang, got in the way of cubicle work.

In some ways, McLean is a throwback, as much an explorer, in the European mindset of the past century's early decades, as a ski mountaineer. His undertaking of skiing the Wasatch's chutes was largely in that mode, albeit within a relatively small geographic range. His idea for *The Chuting Gallery* emerged from a chance encounter with a friend at a Utah Department of Transportation office where McLean was shown aerial photos of the thirty-six avalanche pathways in the mountains surrounding Little Cottonwood Canyon. The DOT was entrusted with digging out the main roads when avalanches shut them down. Photos had been taken so workers would have a more accurate view of where to look for the slides and their runouts. The photos came with slope angles and total vertical feet.

When McLean saw the photos, his mental reaction was "Why not explore and ski these runs?" He recalled in the spring 2018 *Backcountry* article on *The Chuting Gallery* that he took the DOT information and added information he found in a guidebook called *Wasatch Tours*. The irony was that the insights he gleaned from the guidebook were what chutes to *avoid* because of avalanche dangers. The guidebook even had marked such spots with an "A," which McLean interpreted, coincidentally, as the symbol for "anarchy." He recalls: "Between this [the DOT photos] and *Wasatch Tours*' lines of anarchy, there was a five year list of things to ski."

Notching descents that would become entries in *The Chuting Gallery*, McLean continued to explore mountains in Antarctica, Alaska, France, Italy, and all through the American West. But the Wasatch Range in his backyard has been the location of many of his most notable ski runs, and a few of his encounters with avalanches. One that sticks in his mind was a 2004 outing with Bruce Tremper, then the director of the Utah Avalanche Center. The two were in the backcountry on the East Fork of Silver Fork Canyon, near Solitude ski resort. There had been high avalanche danger warnings, McLean recalls, something that Tremper was in a position to know about. Still, the pair were "stomping cornices," testing the stability of the snow below as the broken chunks bounced down the mid-thirty-degree-angle slopes.

McLean dropped in to a not particularly steep slope that Tremper had never seen slide. "Usually when I ski into it, I drop in and do a ski cut," he says of the technique for revealing avalanches. As he completed the diagonal cut across the slope, he started to turn for his line across the mountain in the opposite direction. Then he felt something odd. He first thought he had forgotten to secure his boots, as he was wearing touring boots that free up at the heel for skinning and then lock down at the heel for skiing. Then he realized his feet and knees were being sucked into an avalanche.

"I just lost my control, felt it building up and up. Finally I just Superman-ed down the hill. It was like diving into water." As McLean's body was pulled deeper under he snow, he hit his head on an object. "Then it got dark. I thought, 'I can't believe this.'. . . It was like being shot by a comic book character, Mister Icy or something." What initially seemed unreal, and even worth dismissing as humorous, took a grave turn. "When the snow compacted, I couldn't get a breath. It collapsed my lungs. I'm buried and it's just dark."

Fortunately, McLean could not have been in better company with Tremper close by. "I felt Bruce come and he pulled me out by my legs. I felt like it was two minutes. Bruce said it was like twenty seconds." As he does with many such things, McLean looks back on the encounter as a humbling lesson to be filed away and summoned when traveling in terrain with a high threat of avalanches. "When I teach avalanche classes," he says, "I think maybe the best thing would be for the students to get buried right away."

In recent times, McLean has sometimes added an avalanche survival pack to his backcountry ski gear. When skiing in the Kashmir region of India, he wore a Black Diamond avalanche pack because a lot of the skiing was on "exposed, open terrain." But when he starts thinking of using such gear, he hears the whisper from a part of his brain: "Shouldn't I just avoid that type of terrain?"

In the 1990s, working on his *Chuting Gallery* project showed McLean how few people were tackling vertical descents in the backcountry. But today he sees more and more skiers and boarders willing to repeat his tracks—and add to them. "It grew, and grew and grew," he says of interest in the backcountry, "then about eight years ago it just went"—he gestures with his hand in the air like a rocket ship launching. He attributes a lot of this to the internet and social media. In his early days of pioneering runs, McLean says people thought he was crazy to ski a run like the Y Couloir, which his book describes as hitting "forty [degrees] immediately and never varies more than a few degrees in its entirety." The run is 3,200 feet of sheer vertical.

"Now people are going up the Y Couloir, posting Instagram photos," he says. No doubt McLean could have his pick of young apostles to lead on classic descents, or to help with new projects. A few, like Howell, have become partners, but volunteers are scarce. "It seems like a lot of them are afraid of me," McLean says. "I'd be more than happy to mentor and show what I know. But you have to do it my way."

McLean doesn't wear all his accomplishments on the sleeve of his puffy down jacket. In a profession where the loner soloist, even the high-altitude outlaw, is sometimes quietly revered, McLean almost surreptitiously reveals himself as a leader, as someone just as concerned about his partner as himself.

And he isn't backing down from new challenges. On his calendar are trips to the Swiss Alps and to France's cradle of mountaineering, Chamonix.

But his partners, like Conrad Anker and the late Alex Lowe, are never far from his thoughts. "I think partners are a huge part of backcountry skiing. It's a partner sport. Even though you're doing your own thing, you're always looking out for each other."

18

THE BACKCOUNTRY AT THOUSAND PEAKS RANCH

Going into the backcountry, when you don't have much—or any—experience there requires you to draw on every scintilla of intuition you have; to unflinchingly hew to the guides' body language; to happily, if also disconcertingly, join in survival rituals; to foresee a minefield of hidden dangers; and to ski, if you are blessed, runs that are reflections of unfurling, rapturous liberation.

It's a tricky blend of adult, solemn acumen giving considerable leeway to an unruly, jubilant inner child. My own shift to this state began on a bitterly cold morning at Thousand Peaks Ranch, about an hour's drive outside Park City. I went there to go snowcat skiing with backcountry guide Shannon Finch and her guide partner, Johnny Adolphson. Also aboard the cat would be nine other skiers, people I'd never met. The mix turned out to be wide—a couple of Park City residents, a guy from Woodstock, New York ("the most famous place in New York other than the city itself"), and, among others, a resiliently cheerful Australian woman, a former ski instructor on holiday for a month in the Rocky Mountains.

Thousand Peaks is a sixty-thousand-acre private ranch in the Uinta Range owned by a multigenerational Utah family that leases portions of it out for snowcat skiing, snowmobiling, and even for sets for movies such as the Paramount Network's *Yellowstone*. The exclusive snowcat skiing operation is run by Park City Powder Cats, which takes clients into the mountains to make runs on untracked powder in unpatrolled terrain without heavy avalanche mitigation. What they sell is not so much powder snow as it is the tango with powder snow on the way to transcendence. Or at least that's the ambition.

It's true backcountry skiing without having to sweat the trip to the mountaintops on climbing skins. I had been snowcat skiing twice before, and heli-skiing twice. While a helicopter delivers a certain, jet-powered rush, skiing out of a snowcat is just as good, and, depending on the agility of your group, may allow you to get even more runs in.

I checked in with Finch inside a small log cabin where four snowcats' worth of skiers were banging around the wood floor in ski boots. The room was so full of ski gear—packs, boots, poles, and helmets—it looked like a ski swap sale was taking place. Leaned up against a railing outside were forty or so snowboards and pairs of skis, representing every configuration of contemporary design. With their brightly enameled topsheets, they telegraphed every emotion and physical impulse a skier might evince, from backcountry meditation to slopestyle havoc. They could have been weapons being readied for a special forces mission.

Welcoming me, Finch handed over an avalanche beacon and a liability waiver to sign. As a practical matter, almost everyone signs these forms without reading the fine print. To have signed up for a day in the backcountry and paid your money almost presupposes that you are aware of what you're getting into. You've already accepted the possibility of injury or worse, so why ruin your morning with reading a lot of legalese about it?

Soon all of us would hear a safety briefing, perform an avalanche beacon check, and head out in the snowcats. Beyond the frosted log cabin windows, I could hear the cats' engines begin to roar as their exhaust pipes exhaled a bluish diesel fuel belch into the frigid air. After a brief safety talk, Adolphson and others assigned each of us to a group. How you get picked for one group over another depends partly on what you've filled out in the accident waiver. Skiers are asked to rate their level of expertise, and along with age and other details, groups are assembled and placed with specific guides.

Backcountry snowcat guides typically work in pairs. Heli-ski guides, flying clients into terrain that is adjacent to a resort, often pair up with a ski patrol member who acts as the guide's temporary partner and point person for injuries. On this day Adolphson would ski lead guide for our group with Finch skiing tail gunner. Outside, before boarding the snowcat, Adolphson waved his avalanche beacon over each of ours, the ensuing chirp like an alarm system on a door opening wide into the backcountry.

GATHERED INSIDE THE CAT, WE settled into our seats and ventured quick introductions. Because people have paid a lot of money for snowcat or heli-skiing, there is a faint illusion of invulnerability, or at least of a special protection bestowed by the presence of guides. As experienced professionals, they go through a checklist to minimize danger and fallout from accidents. I was once on a snowcat headed into the backcountry and a guide asked if any doctors were on board. A guy raised his hand and reported: "I'm a veterinarian."

"Even better," the guide wryly responded.

Beyond the guides' regimen of protocols, kinship on a snowcat is an odd and evolving dynamic. Early on, everyone is immersed in the morning's mental fog. There are private glimmers of hope about what the peaks and snowpack hold, but until that terrain comes into view, it's a hazy dream.

No doubt, people are quietly sizing each other up. Surreptitious glances are made at each other's gear. Certain types of skis, their lengths, widths, bindings, and make, tell a story. Longer skis and wider ones—say, 112 millimeters in width and up—are true powder skis meant for days after solid snowfalls, or "freshies." Owning a pair usually means the skier has a "quiver" of skis meant for different conditions. Free-heel bindings—telemark style—or alpine touring bindings (AT for short) indicate a skier has likely done time on climbing skins in the backcountry. Standard, alpine-style downhill skis, with permanently locked bindings, often means the skier is mostly limited to resort skiing. Other gear—ski shells, pants, backpacks, helmets, even gloves—may give hints at the dues a skier has paid in the backcountry. The one piece of gear that is a pure giveaway of extensive out-of-bounds experience is personal ownership of an avalanche beacon.

On our way to the first run, Finch said we would start the morning with "something mellow." A few minutes later, I stood at the top of a fairly steep slope that plunged down into a cross-hatch of tightly spaced trees. My experience "skiing trees," as some enthusiasts refer to it, is limited. It's not something you usually encounter on groomed runs in resorts, so you have to consciously select runs into terrain shadowed by trees. It's a strategy many skiers will pursue in storms because snowfall is partially blocked by tree branches and, without snowflakes clotting on your goggles, visibility is better. Absent sunlight on an open run, you ski in flat light, which reduces your ability to see the ground and fall line down the mountain. With no tree trunk or branch shadows on the slope, all the slope's bumps, dips, and sudden drop-offs coalesce into one long, hypnotic expanse of white.

Despite my misgivings, I readied myself for the tree run. Thinking back to another day of cat skiing in the Wasatch, I recalled hearing the guide say we were headed to the Enchanted Forest. I mistakenly assumed the name was born of some psychological or even spiritual dividend the skier received at the end of a run. It turned out the run had been anointed such for its perfect spacing of aspen trees, trees so strategically positioned by nature that the glade looked more like a skier's Disneyland exhibit than a wooded obstacle course. Even the shadows cast by the bleached trunks looked as if they'd been designed by a lighting director.

But the run that stood before me on this morning in the Uintas possessed no air of enchantment. It looked foreboding and bumpy, and offered a thick copse of trees as welcoming as a mugger. I waited until all the other skiers had dropped in. Now it was just me and Finch. I told her of my limited experience skiing trees. She said something encouraging about the morning's half light between trees. "Focus on the space between them, not the trees," she advised.

Finally, I dropped in. And promptly caught an edge and fell. I bounced up quickly, aimed for some daylight below, and raced past trunks and branches until I made a sharp turn to slow myself. Then I fell again. As before, I was up quickly. But I waited for Finch, who was of course skiing beautifully.

As she pulled up beside me, she asked, "How you feeling?"

"Sketchy," I responded.

"Well, just take it easy," she comforted.

Skiing into the gully below, I bore right down a narrow track to the snowcat waiting ahead, its engine humming as skiers dropped off their gear into the metal basket anchored to its large quarter panel. Once inside the cabin, I took some deep breaths. I had not slept well the night before and had had only a single cup of coffee for breakfast. It was still bitter cold outside, but I could see the sun was gaining altitude and knew the day would warm considerably. It seemed a good idea to skip a run and reset my focus. The snowcat growled into gear and began its trek toward a high ridgeline in the outlying peaks.

BEING IN THE BACKCOUNTRY IS, in my limited experience, an occasion for submitting oneself to a volley of emotions and physical sensations. There is the disappointment in finding one's skills lacking—my morning's run-in with the trees—and there is the exultation that comes from living for a day in an otherworldly, wild, sometimes hostile domain, which is often also stupefyingly beautiful. Along with all these sensations there are the nagging aspirations of fulfillment. And that means the anticipation of having almost indescribable, joyous runs.

The late novelist James Salter, in an essay titled "The Skier's Life," observed that "skiing, like dancing, seeks to be admired." The catch, as he astutely observed, is that just when you feel your skis getting dialed into the snow, when the turns seem to instinctively find a groove for themselves, you see someone off to the side performing turns of eternal grace and realize he or she is, as Salter put it, "the skier you will never be."

A day in the backcountry can end in transcendence, but to get there one may have to endure some humiliation. Fortunately for me, the former was what the rest of my day in the Uintas held. As the sun broke above the peaks, the cat driver invited me to sit in the cab with him. Seated within the glass bubble, I was rewarded with a view of the Uintas that I would have paid good money for, forget the skiing. With a vantage higher up than I'd had from skis at snow level, I saw the mountains at Thousand Peaks fan out like a massive sports stadium, its bleachers vast, Roman Colosseum–like oval tiers climbing skyward, drenched in white. The telemetry of the slopes seemed of a magnitude beamed from space. As the risen sun flooded over the snow, a prismatic glow arose from the steep pitches that flowed into valleys below.

"I have the best office view of anyone I know," the cat driver proclaimed.

He was right. Like that of a helicopter, the cab was all tempered glass but squared off instead of bulbous. Part monster truck, part moon rover, the snowcat is called a PistenBully, the undercarriage of a tank mounted with a twelve-seat steel box on back. The vehicle bears more resemblance to a robotic Transformer than a bus. With a 350-horsepower diesel in its underbelly, and stainless-steel treads running its length, the cat chugs along at ten or twelve miles an hour. After reaching a designated mountain peak and discharging its passengers, the PistenBully has the ability to turnaround within its own footprint, which makes the cat feel, as it rotates, as if it's consuming itself. And then, suddenly, a new panoramic alpine vista comes into focus.

Other than a helicopter, which is a giant, mechanical cousin to the dragonfly, and an F-16 Hornet fighter jet, which a Blue Angel pilot once let me steer into a complete roll above the Atlantic Ocean, I can think of no contraption as whimsically assembled as a PistenBully snowcat. Sitting in the cab of one, grinding your way up a forty-degree mountainside on a ten-foot-wide ridge is like donning Robert Downey Jr.'s Ironman suit. It's a machine built for Armageddon. As it turns out, it is perfect for backcountry skiing.

Inside the cockpit, I felt the pathfinding machinery of the PistenBully paw its way down to pick up the other skiers. The sun above us was beginning to project warmth. The sky was the color of the Aegean, only fully scintillated with fleets of solar javelins. As the group of skiers gathered around the PistenBully to deposit their skis in the carry basket, I wondered what had compelled them to sign on for a day in the backcountry. The group was mostly composed of middle-aged skiers, a few of whom carried a ski bum air about them, glove rips duct-taped to keep the snow out. Others looked like prosperous professionals from Park City or Salt Lake.

Motivations to put oneself in the backcountry, even when accompanied by guides and riding inside a VIP bus on tank treads, are as varied as personalities. Some are there for untracked powder, others to get out of their ski resort comfort zone, and still others to chase an ethereal state that is part of the proposition promised to those who venture into the out-of-bounds. "Find freedom in the backcountry," read an email I received one winter morning from a Park City gear company, Backcountry.com. "You get up at the crack of dawn in search of those untouched lines. Why? Because nothing goes together quite like fresh powder and backcountry touring."

Even when a group of backcountry skiers has come together with no previous knowledge of each other, there are protocols that tend to play out. Usually they have to do with a kind of athletic craftsmanship, a subtle language of communication that emanates from the body and through brief verbalized remarks and a pecking order that, like water, seeks its own level. A skier's regard for craftsmanship reveals itself in things like the choice of ski models and the materials used in them, wood and titanium and lightweight, expensive carbon fiber woven into the skis like black magic. It's an atelier mindset expert skiers gravitate to as they select skis for one reason or another.

Like competitive sailors, backcountry skiers tend to be a laconic bunch. They'd prefer their bodies to broadcast as much or more than their mouths. Through experience in the mountains, they've tapped into a stratum of information sharing that, like a high-pitched radio frequency, only they and certain others can hear. If you aren't tuned into it, you are always a few yards off the pace, the last person to click into your bindings because you missed the signals. It's as much a language of eyes as of ears.

Pride of ability discloses itself spontaneously when a guide gives the go-ahead for the first skiers to drop into a run. Like cars meeting at an intersection with four-way stop signs, there may be a bit of hesitation, but inevitably someone grabs the initiative and goes first. The next three skiers swiftly self-select themselves to follow. And from there, for the remaining three or four skiers, it's a mental game of hesitation hiding behind manners. Once that order has been established, it tends to repeat itself for nearly every run.

AS MUCH AS I'D ENJOYED the view from the PistenBully cab, it was time to get back on the snow. We ascended to a peak overlooking a wide bowl with no obstacles in sight, just unfurling vertical planes of untouched snow. The hesitation I'd felt earlier had melted away in the mid-morning sun. Once I was secure in my bindings, I skied a traverse to my left, and Finch followed behind me. At the top I paused for a few seconds as she drew to a stop maybe a dozen feet away. Then we both dropped in.

Immediately, I found rhythm in my turns, my pole plants soft jabs, my knees connecting with the jackrabbit bounce that fed off the trampoline snap of the snow. Turn after turn, I shed my deliberations with my skis, and fell into a kind of meditative state where speed and just the tiniest increments of effort, more like thoughts than muscle, governed direction changes. As I finished every turn, I could see, out of the corner of my uphill eye, the shadow of the rooster tail of snow flung into the air by the tails of my skis. As if in slow motion, the snow was landing with its own subtle hush, as if each flake was like a fingertip on the earth, a slow, white narcosis enveloping the mountain behind me.

In the runout below I rounded up near the PistenBully, Finch just feet away from me. We looked up toward the peak where we'd initiated the run. "Look at our turns," she said gleefully.

Like something from a physics text, our turns were two mirror images, precisely matched, not so much something drawn in the snow but carvings suspended in time, the frozen currents mapped as if visible contrails left by the feathers of swerving swallowtails. All afternoon, runs like this ribboned out behind me in whirls of white. The Uinta peaks stood, from valley to valley, like an array of gargantuan bleached conch shells.

We paused for lunch in a place called Cold Smoke Canyon. It was a sun-filled amphitheater, the picnic spread out before us. No other human presence was evident, no other movement of conveyance, no other sound. Surrounding us on all sides was Craig Gordon's ubiquitous "spooky snowpack." "All it takes is that one chink in the snow that propagates a crack," he told me when talking about his experience in the Uintas, "and it communicates that insult to the rest of the slope. And the rest becomes history."

Finch had spoken to me of Gordon's partnership with the Powder Cat guides. He frequently conducted avalanche awareness sessions at Thousand Peaks Ranch, one of which I later attended where he introduced me to the Stairmaster of all Stairmasters—climbing skins on my skis going up a steep mountainside. Eventually we broke through a stand of trees onto an open mountain face, snow gleaming and immaculate in the sun as if it had been sitting there since the dawn of time. And then I stripped off my climbing skins, locked down my bindings, and dropped into my first turn on skiing's Promised Land.

Gordon, said Finch, often phoned the guides for firsthand reports of snowstorms in progress. "During storm cycles we talk maybe twice a day," he told me. The line of communication with the guides had been a great boon to Gordon's avalanche advisories on the Uintas.

"You would have an idea of what was happening," he said of his reports in years before input from the Powder Cat guides. "But prior to that help you would just throw a dart at the wall." A mutual sense of respect had grown between Gordon and the guides, Finch said, and Gordon echoed it. "In the past decade, they've really come into their own," he told me. "It's a world-class operation. They have a very respected operation with very respected guides."

The partnership had become so tight, Gordon said, that Park City Powder Cats had agreed to occasionally donate the use of two powder cats to Gordon and the Utah Avalanche Center. The avalanche center could then raise money for operations by selling seats on the cats. "It's like someone going out and picking out a gift that really means something to you," Gordon said.

We packed up the lunch leftovers while Finch and her partner aimed the PistenBully for a ridgeline a little higher than what we had been skiing. By the time we had climbed to the drop-off spot, the sun was on its downward arc, still full over the peaks, but the light angling just above thousands of mountain acres deeply buried beneath snow crystals that seemed to vibrate.

We stood on the ridge, eyeing the steep pitch down. Finch had been given the job of making a ski cut across the slope to test the snowpack for avalanching. I watched her make the ski cut traverse toward the sun. The wake of her skis left narrow trenches that filled with light like water bouncing over rocks in a shallow stream. And then, on her free-heel telemark skis, Finch made a slow curve to her left and dropped into the steepest part of the mountain's face.

It was as if she had transformed her entire physical being into that of a raptor. Slowly, beginning at six or seven thousand feet of altitude, she was in her turn earthward. Then, as her body angled sharply forward, her arms tucked in close to her ribs as if she was folding wings back for aerodynamic perfection. Within seconds, her speed seemed to have doubled. She began to stitch a slim series of turns.

From where I stood, I could see Finch's lead ski reach down the precipitous face as her knee and thigh followed in a stretch of pure physiognomic defiance. The tips of her ski poles were like claws barely piercing the air. Behind her silhouette, the afternoon sun captured each arc like advancing frames shot from a camera's motorized shutter. *Click, click, click* . . . .

And then, far below, Shannon Finch was gone. An invisible slipstream of air molecules seemed to stir over the slope where her body had just zoomed past. I had seen her disappear faster than any wind combing the high spires of Thousand Peaks Ranch.

19

THE HIDDEN STREAM AVALANCHE: GARY ASHURST

Just six miles outside the center of Ketchum, Idaho, a narrow, shady road leads to a paradisiacal patch of mountain land where Gary Ashurst, home from a hard day on a construction job, sought to rinse the week from his skin by plunging into the clear, cold riffles and pools of the Big Wood River, a wandering valley stream that licks at the shoreline of his backyard. It was a ritual for Ashurst on a hot Friday in July, right up there with spring fly fishing and summer hiking on the trails of the Sawtooth Range peaks that hover above his home perched at 6,500 feet.

But if Ashurst's life in the mountains—ranges in Idaho, France, Italy, Morocco, South America, and many other alpine ridgelines—is one filled with many rituals, there is none more important to him than when snow begins to fly and he clicks his boots into ski bindings and begins a steep descent from a peak. More often than not, his runs border on freefalls, oftentimes after rappels allow him to gain access to never-skied terrain.

Ashurst, now sixty-three, has lived many of his adult years in Ketchum, skied every meter of the Sun Valley ski resort, and a lion's share of the Sawtooth's backcountry reaches, much of it solo. As a backcountry guide with nearly forty years' experience, he is one of the few Americans ever to earn official ski guide certification in Europe. He is known around the world, among a cultish group of ski mountaineers, as a man who forged legendary first descents on peaks and down chutes that form the exoskeleton of La Meije, the massive mountain in France that seems to almost threaten the tiny village of La Grave that kneels at its feet.

Winter after winter, over thirteen seasons, Ashurst guided in La Grave, putting up new routes, introducing clients to ski mountaineering adventures, and establishing a reputation as a guide who could urge his charges to confidence levels—and down pitches—they never before imagined possible. He packed so much adventure into his years in La Grave, it constituted a lifetime of its own. It's where he met his wife, Julie Johnson and where his daughter, Alagna—named after an historic Italian ski village—spent most of her childhood. He built a home in La Grave where Julie presided over weekly dinners attended by skiers and boarders from across the globe. To the extent any American can be embraced by the French as one of their own, Ashurst, through his skills, his soft, confiding smile, and a calm domination of seemingly impossible ski descents, became part of French ski mountaineering circles.

Given his years on La Meije, a 13,071-foot massif, close calls weren't unknown to Ashurst. But they were just that—close. One day that all changed. On a mountainside above the border between France and Italy, on a sunny day in February 2006, Ashurst saw a ski run unfold like no other he'd experienced. As he and two friends stood on a peak, preparing to drop in, a French border gendarme glanced up and spotted the men, two thousand feet above him. The gendarme saw the first small figure leave a puff of snow in his wake—Ashurst stitching symmetrical furrows in the powder, turn after turn. Then a second figure pitched forward from the ridge—fifty-six-year-old John Seigle of Aspen, Colorado, a longtime client and friend of Ashurst's.

What happened next erupted before the gendarme's eyes like a long, simmering secret suddenly released into the world in a ferocious white cloud. More than twelve years after that day, Ashurst sat in his backyard by the Big Wood River, sipping a beer. Raising his arms above his head to stretch, he revealed a long scar across his left bicep, the fleshy ghost of surgical sutures, a remnant of that day that went so rapidly from an affirmation of glory in the mountains to the physical margins of survival and death.

UNLIKE MANY SKI PROFESSIONALS WHO are introduced to the slopes as small children barely after learning to walk, Gary Ashurst was nearly into his teens before he linked turns on anything resembling a true mountainside. For him it was more a matter of overcoming the California geography than a lack of desire. Growing up in the foothills of San Bernardino, in the shadow of the San Gabriel

Mountains, Ashurst knew where the snow was. But with parents who didn't ski, he lacked any mobility to get himself there.

Even as young as five, Ashurst recalls having a longing to strap skis to his feet. "I would say I was bitten by skiing at age five or six watching it on TV. Whenever it was on the *Wide World of Sports*, I was glued to the TV." When he was eleven, an uncle who skied gave Ashurst an old pair of his skis. "They were hardwood skis with bear trap bindings, cables," he recalls. It turned out there was a learner's slope not far away called Snow Valley. The "Valley" part should have been a tip off; the hill only had about a thousand feet of vertical, if that.

"I'd hike up and go straight down 'til I crashed," Ashurst says. This punishing apprenticeship went on until he managed to acquire a pair of skis with metal edges and a pair of double lace-up boots, the latter found by his mother at a garage sale. One of his sisters had a friend whose family skied, and Ashurst ingratiated himself into ski trips to Snow Summit ski area at Big Bear, a progression that put him on "a big mountain, two thousand vertical feet."

When he turned sixteen, Ashurst took his flirtations with the snow to a whole new, far more serious level. "Once I had a driver's license, it was game on." He enrolled in summer school, not because he was particularly academic but because the extra classes allowed him to take an abbreviated schedule in the fall and winter, and leave school by noon. Making a beeline for Big Bear, Ashurst and a friend would reach the mountain and ski until closing at 10:00 p.m.

If Ashurst had been clever in creating ski time by juggling his schoolwork, he was even more so when it came to getting lift access for free. He befriended the lift operators by giving them marijuana joints; he would palm one to a liftie for the afternoon, and when the night operator came on at 5:00 p.m., he would fish another joint out of his pocket and ride on into the night. Ashurst's diligence began paying off as he became more comfortable with speed (later in life he would coach clients, saying, "When skiing powder, speed is your friend"); his newer skis' metal edges began to bite deeper into hardpack snow as his turns responded to more adept body language.

Soon he was training with Big Bear's racers, running gates at night. The advent of freestyle competitions came along: balletic skiing that included moguls and explosive, gymnastic jumps at the bottom of the competitors' course. Ashurst immersed himself in all of it, with the goal of improving his all-around skiing abilities—techniques that later translated to more agile, adept backcountry skiing. But his aspirations were still tied to a shoestring: Ashurst shared a friend's pair of Lange ski racer's boots so that when the friend finished a race, he had to quickly unbuckle the boots and pass them to Ashurst.

When Ashurst graduated from high school, he was admitted to Southern Oregon University, his primary interest not scholastics but joining the school's ski racing team. "That's when I found out there were real racers," he recalls of his introduction to the team. His dedication and discipline began to get noticed. Scott USA, the ski equipment manufacturer, offered to sponsor him and provide him with boots, goggles, and poles. As Ashurst's skiing exploits were advancing fluidly, his college studies were frosting over, but he managed to get a degree. At twenty-one he found himself contemplating his next move. By chance, a Scott USA rep learned he was trying to decide what to do.

"Come to Sun Valley," he told Ashurst. "You can have a job at Scott."

Arriving in Ketchum, Ashurst found both the cradle of American skiing, amplified by the legacies of Hollywood and Hemingway, and the big mountain allure of the Sawtooth Range. He discovered one other thing in the surrounding mountains—incomparable powder skiing.

At the Scott ski boot factory, he learned there were three shifts per day, including one that started at 3:00 p.m. As with his high school schedule, Ashurst figured he could work from 3:00 until near midnight, grab some sleep, and then hit the mountain at sunrise. "It was perfect for skiing," he remembers. "And I continued to do the races here in Sun Valley."

Over time, Ashurst realized that while his passion for skiing and chasing powder in the backcountry had deepened, his focus on racing had hit a plateau. He'd been accepted into the Sun Valley tribe of families composed of generations of venerated skiers, racing champions, and hardcore ski bums. He started coaching young racers on the D-Team, eight- to eleven-year-old nascent racers. Having left behind his Scott factory job, Ashurst and a friend founded a cabinet and woodworking business.

For the next decade, Ashurst sank deeper roots in Ketchum and the Wood Valley. The construction business flourished, and Ashurst hunted down powder stashes that in the 1980s were still the haunts of only the most devoted backcountry skiers. But the freedom he saw in lives of ski wanderers, those patrollers and instructors who came to Sun Valley and left with the seasons, did not escape his attention. In 1989 he sold his share of the construction business to his partner and took a year off to travel. He bought a piece of land outside Ketchum, but he left and moved to Maui for three months to wind surf. He then drove his van down the California coast and into the Baja Peninsula so he could windsurf and surf the waves of the Pacific.

Eventually, on the horizon, past the blue wave peaks, Ashurst sensed a white intimation of winter.

ASHURST HAD A FRIEND WHO had been living in Chamonix, France, during ski season. When they connected, Ashurst asked the friend if he planned to return to the stronghold of European alpinism and steep skiing mastery. "He said, 'Yep, I'm going to France this winter.' He said, 'Definitely go to Chamonix, but they opened up this place, La Grave. It's like Chamonix but with no people.'" Early that winter, Ashurst found himself driving a Volkswagen Vanagon across England, headed into France. All the way through the French countryside and into the Hautes-Alpes of the southeastern region of the country, he saw snow falling. On December 1 he pulled into the tiny twelfth-century hamlet of La Grave.

Ashurst promptly ducked into a café to gather intel about skiing. To his dismay, he was told that the telepherique, the single, large cable car that ferries skiers up La Meije, was not scheduled to operate until January 24. As promising as the setting of La Grave looked to Ashurst—intimate stone houses clustered along terraced slopes that bowed before La Meije's broad shoulders—he had come to France to ski, not sit in a café waiting for more than a month.

By a fortunate coincidence, Ashurst knew some Sun Valley ski patrollers who had traveled to the nearby ski resort of Serre Chevalier in an international patroller exchange. He drove there and skied for nearly a week. With the legends of Chamonix a continuing echo in his mind, Ashurst made his way to the storied town. There he skied for a week and began assembling gear for his return to La Meije. Buying *randonee* equipment (European free-heel skis), crampons, and rope, Ashurst headed back to Serre Chevalier, where he skied for a second week, this time testing his new gear.

La Meije awaited. On January 24, 1990, Ashurst rode the telepherique up to the Girose Glacier that constituted the skier drop-off spot. Stepping into his bindings, he took a long look around the snowy panorama before him. Almost no other skiers were present. Then he started down. "It was incredible," he recalls of his early runs. "We've hit it, jackpot here. . . . The first day I went up to ski, I was going up thinking it would be mellow. But after my first run, I went down to get my fifty-meter rope."

If La Grave was a rumor destination for serious skiers, the hushed reports about it and the sudden appearance in the town's cafés of vagabonds toting backpacks and ski bags was a source of mystery to locals who'd been left out of the loop. Most of them were farmers who didn't ski, and the young, hipster skiers were initially a source of mild amusement to them. Familiar with tourists who came in the summer from Grenoble and Geneva to take in the cool mountain temperatures and hike the lush valley trails, the locals were totally unprepared for these winter visitors. After all, La Grave was not a ski resort like Serre Chevalier; there was only a single lift, no ski patrol, no marked trails, no avalanche mitigation. It was more than the backcountry. In winter, it was an alien planet on which travelers from another galaxy had suddenly landed.

"We were an anomaly," Ashurst recalls of those early days. "We had come there specifically to ski. . . . It was difficult to find a place to live. They rented places out in the summer. People didn't come there to visit in the winter." Soon he met the four or five other American skiers who had trickled into the village, compatriots from Colorado and Sun Valley. "We kind of had a heyday," Ashurst remembers. "A lot of this stuff hadn't been skied. We went and discovered runs." With no ski run map to follow, he and his crew got inventive in reconnoitering pitches and couloirs and ways to link them. "A friend with a parasail would glide above and report on them from the air."

As the Americans and a handful of other arrivals began to pioneer more lines on La Meije, all the exuberant, often audacious—not to mention death-defying—descents did not go unnoticed by the few French guides who worked on the mountain. Ashurst says they "started following our routes." The French guides weren't the only ones following Ashurst. Charlie Johnson, a Sun Valley ski patroller, was one of the exchange patrollers living in France. His sister, Julie Johnson, had decided to fly over to ski with her brother, but when she got to the resort the snow was less than appealing. So Charlie, aware La Grave had received a lot more snow, advised her: "Give Ashurst a call." Of the experience, Julie said: "That was the first time I'd skied La Grave, and it snowed the night before. It was just beautiful."

The following year, the winter of 1990–91, Ashurst returned to La Grave and met two Swedish guides, one of them Pelle Lang, who became a lifelong friend. At La Chaumine, a classic alpine chalet perched above the center of La Grave,

Lang had started a winter skiers lodge. After skiing with Ashurst, Lang offered him an opportunity to guide out of the lodge.

"At the time," Ashurst recalls, "I didn't have any certifications. I was very lucky. I wouldn't have been able to do that in Chamonix." In fact, in order for Ashurst to lead clients from the top of the telepherique across the Girose Glacier, which was pockmarked with hidden crevasses, he had to pair up with a French guide who had glacier travel certification. But Ashurst earned the respect of the French guides, and in 1996 he passed an eight-day exam in Switzerland and formally added an International Mountain Guides Association (IMGA) certification to his guiding credentials. The tests were no lark for him. "There was a pretty significant amount of hazing. They tried to rattle you. My MO was, if I'm going to fail, I'm going to fail on my own, not because of them."

By the late 1990s, La Grave was no longer a word-of-mouth tip in a darkened skiers' bar. It appeared in ski movies such as *Snow Drifters* and in short but mystical magazine descriptions that created more allure than if they had been fully blown feature stories such as famous American and European resorts received. Ashurst saw the weekly arrivals of what he called visiting "ski bums" go from about 30 a week to nearly 150. "There was a different mentality in the people who came there to ski for the season, living there and skiing as opposed to those who came to just visit Chamonix," he says. In Chamonix, at the end of the day, there was as much outlaw behavior in the bars as on the piste. Not only was public partying absent in La Grave, there were almost no places to host any bouts of communal drunkenness.

A dearth of watering holes wasn't the only drawback for skiers less willing to commit to the winter vagaries of La Meije. "La Grave was a hard place to live," Ashurst says. "Everything is difficult. It's steep. . . . The snow removal wasn't regular. . . . It was difficult to get up and down the hill." By the "hill," he isn't referring to La Meije. From the center of town, the telepherique is less than a five-minute walk, even if it's through a fresh snowfall. But in 1993, Ashurst had moved up the hillside to be closer to La Chaumine. Sometimes after a big snow, cars could not navigate the unplowed, switch-backed road; the only way to reach the lift was to ski the heavily snow drifted road.

As splendid as La Meije could be on a morning when a foot of new snow was draped under the first rays of a rising sun, the mountain and valley had a split personality. On some days—or some weeks—storms blew in with fifty to sixty mile-an-hour winds, and under such heavy snowfall the entire valley seemed to be drowning in snowdrifts. “You are hunkered in there,” recalls Julie Johnson. “We would have three- to four-day storms that would, with the winds, leave meters of snow.” Winter residents learned, Johnson says, to have enough food on hand to weather the blizzards. There were other consequences from such storms—road closures in and out of town because of avalanche dangers.

By 1991, Johnson had returned to La Grave. “I basically went as a ski bum. And Gary was figuring it all out.” The following summer, Ashurst and Johnson were a couple. They rented out their American homes and, living out of a motor home, traveled to Mexico, hung out on the beaches, then went rock climbing at Red Rocks in southern Nevada, followed by camping and climbing at Idaho's City of Rocks. “We were like living the retired life and somehow getting away with it,” Johnson says of that time. Arriving in La Grave for the winter of 1996–97, they sought to improve their living situation to better endure tough winter conditions. They bought an unfinished condo near La Chaumine, basically a concrete shell that Ashurst built out, even installing a washer and dryer, a rarity in the valley. There was another reason for all the sudden focus on domesticity. “It became apparent I was going to have a child,” Johnson says. “And Gary became more businesslike.” Their daughter, Alagna, was born before New Year's.

The Ashursts became a focal couple in La Grave's ski scene, both as residents within the circle of professional guides and as hosts to the clients who continued to return to ski with Ashurst, who by now had established his own private guiding operation. Near the end of ski weeks, Johnson put on celebratory dinners for clients at the Ashursts' home. Sometimes she'd ski with her husband and his clients, although she admits being partial to skiing with “the ski bum group.”

Johnson came honestly by her penchant for skiing with hardcore mountaineers. Her parents were both ski instructors at Crystal Mountain ski area in Washington state. Her grandparents had skied on Mount Rainier when the notion of ski resorts wasn't even a gleam in a developer's eye. Johnson's mother had died when she was young, so being in the mountains brought back happy memories of a time when the family would gather in a rustic cabin near Crystal Mountain. The cabin was so bare bones, Johnson recalls, the most essential thing was not food and water but matches for starting fires for warmth.

Despite living in La Grave for so many winter seasons, many of them arduous, Johnson says she never lost her original sense of reverence for runs down La Meije. “Being with someone in the French Alps and extreme skiing, it seemed normal to me," she says. She doesn't take for granted certain moments on La Meije. "The most amazing ski runs ever," she remembers. "If you're standing on the edge of a cliff and your ski tips are hanging over and it's a two thousand to three thousand foot drop . . . it's a moment of awe, and it's got your attention."

The day that stands out in her memory more than any other on La Meije was when she and her husband skied the Pan du Redeau, a three-thousand-foot couloir with an entry that can only be reached via a climbing rope rappel. Once deposited into the top of the couloir, the pitch beneath the tips of your skis falls away at an angle of more than fifty degrees. Vertigo is not an option. The run is branded a "no fall zone" or, in less euphemistic words, a run where "you fall, you die." She remembers the thrill: "It had just snowed, and you go a hundred, two hundred turns down in powder snow, untracked. Where else are you going to get that? And I got to do it with Gary. It was just the two of us."

If her husband hasn't exactly mentally left La Grave behind, Johnson says the ski experiences with him in France have translated for her to the Sawtooth Range backcountry. "Now we go up into the hills in Idaho, and I will follow him anywhere. We have that connection. . . . Step off the radar. I think that's what it is."

TWO YEARS PRIOR TO ALAGNA'S birth in 1996, as Ashurst was beginning to see clear signs of his business taking off, he got an inquiry from a skier in Utah who wanted to come to La Grave for a week. Ashurst took the booking.

The skier's name was Andrew McLean. "It was obvious that Andrew was not cut from the same cloth as most skiers," Ashurst remembers of his first impressions of McLean's abilities. The days skiing with McLean raced by as the two men, and McLean's wife at that time, devoured descents. "We had just barely scratched the surface," Ashurst remembers.

At the end of the week, the Utah skier told Ashurst he wanted to come back the following winter. Ashurst decided the two had to put in more time on La Meije together, so he offered McLean a job with him as a guide. Thus began a two-year professional relationship between them. By the late weeks of the 1996 spring winter season, Ashurst and McLean had skied together so much they had developed an intrinsic mountaineering relationship that seemed as though it could win them any challenge La Meije had to offer.

"One day," Ashurst recalls, "he said, 'Is there anything on your hit list you want to do?' I knew exactly what it was." Standing on another peak many months earlier, Ashurst had eyed the run through binoculars, tracing as much of the line as he could see, wondering about the steepness and conditions of the sections hidden within the walls of a couloir. The summer before, he had even hiked up to try to scout out the pitches, which he deemed to be angled as much as fifty-eight degrees at the top, step-off-a-skyscraper steep.

He eventually concluded that the run, which no one had ever undertaken, had "double terminating" sections. This meant ski partners would have to each rappel down two rock faces in order to complete the three sections of the run. The rappels were well below a convex drop in the peak from which one could not see over in order to judge the length of rope needed.

Always game for adventure, McLean not only agreed to team with Ashurst on the first descent, he also memorialized it in a detailed sketch. First naming the run "The Big Bang," McLean noted it would be "Thirty-two hundred feet to valley floor"—assuming they made the rappels and didn't fall to their deaths. In his last footnote, McLean observed: "Big exposure!"—the word "big" underlined twice.

On their first attempt of the run, the two men dropped in and skied the first three hundred feet to where the couloir narrowed to about six feet and the first rappel was needed. From that vantage point, they could see they had underestimated the roped descent. "We stopped and took inventory of equipment," Ashurst recalls. "We made the correct decision we didn't have enough rope. It would have ended up being a rescue."

The next day, they returned, again skied the first pitch, and set, as McLean described them, "bomber anchors" at the top of the first rappel. When the time came, McLean went first. Only one man could ski a pitch at a time because of avalanche concerns and the need for a sentry over the rappels. It took them hours to achieve the first two sections of the run. McLean, in his sketch, and Ashurst recalling the feat, estimate the first rappel was between twenty-five and fifty meters and the second was seventy meters.

When the rappels were complete, they packed away their ropes. Then they looked toward the valley below where the run terminated in an "alluvial fan," as Ashurst describes it. A small pinnacle stood to their left. The two men, bindings locked, kicked forward and skied nearly three thousand feet to the bottom.

That night, recalling the run, McLean said he didn't feel like "The Big Bang" quite captured the line. On his sketch he had made a small note of a rock feature near the top he labeled "Gargoyle." Ashurst recalls: "We talked about it, threw names out. He said how about 'The Gargoyle'? And that stuck. Now it's a test piece for young, aspiring big mountain skiers."

In many ways the descent of The Gargoyle captures a lot about the myth making (and the legend creation) that goes on in La Grave. When young, cocky skiers and boarders arrive in La Grave, Ashurst says, "It doesn't take long before they say, 'What's the hardest thing here?' And people say: 'The Gargoyle.'"

JON SEIGLE WAS A FIFTY-SIX-YEAR-OLD resident of Aspen, Colorado, who was a member of a Wednesday morning Aspen Mountain ski group that had formed an opinion of him as "by far the most safety conscious." There was visible confirmation of that view: Seigle always wore a helmet in a time, the early 2000s, when skier and snowboarder helmets, viewed by some as nerdy and unnecessary, were just beginning to be adopted. He sometimes skied with a device called an Avalung, which was Black Diamond's forerunner to the JetForce avalanche rescue packs; the Avalung was designed to allow a buried skier to breathe in through an intake valve near the chest while exhaled air is directed through a tube to lower down, thus delaying death by carbon dioxide poisoning.

"In all our minds he was the most unlikely person in our group to even be hurt," one of the group's members told the *Aspen Times*. A member of the group recalled Seigle as always being the first person on the mountain, often found at the peak when it was still dark, and the sun was about to rise.

"Why so early?" the member once asked Seigle.

"Because I want to eat breakfast with my girls," the answer came.

In a town like Aspen, known for residents who can sometimes put on airs, Seigle was familiar to many as an unassuming man, gracious in his relationships with on-mountain patrollers and ski lift operators, the lifties who are often classed with low-level workers like dishwashers and maintenance crews. "He knew all the lift operators on Aspen Mountain by name," one of Seigle's group recalled to the newspaper, "and would bring gifts to them on the last day of the season."

While Seigle mostly stuck to skiing the mountains surrounding Aspen, he had by all accounts achieved such expert status that he inevitably decided to make some ski trips to Europe. Where he chose to repeatedly go was La Grave. But even in his early trips to La Grave, Seigle is said to have received "assurances from a La Grave guide . . . that they wouldn't do anything he wasn't comfortable with."

That guide was Gary Ashurst. "He'd been coming to La Grave every year for four or five years with another friend, Gary Sawyer," Ashurst recalls. "They would come over and stay with me and sometimes ski with me and sometimes do their own thing." Of Seigle, Ashurst says: "He was a great skier, actually quite conservative."

In the latter part of February 2006, Ashurst had been skiing in Val d'Isere, France, for a week when he returned to La Grave. Seigle and Sawyer had arrived in town, and Ashurst met up with them. The previous year, Ashurst and Julie Johnson had sold the condominium they had built out and furnished, and Ashurst was still sorting through stored possessions. But he liked Seigle and Sawyer, considering them among his more favorite clients. "I said, 'If I'm skiing, I'll ski with you.'"

The 2006 Winter Olympics in Torino, Italy, were going on, and when Ashurst and others watched the events on TV they could see snow falling in the Italian Alps. But La Grave had been hit with an unusual stretch of no snowfall. In addition, the avalanche risk had risen, so much so that the telepherique transporting skiers up La Meije had been shut down.

After some discussion, Ashurst, Seigle, and Sawyer decided to drive to Montgenèvre, a ski area on the border of France and Italy. It was familiar terrain to Ashurst, and he and the others found fresh snow awaiting them. "The first day, it was unconsolidated [snow]," Ashurst recalls. "I'd been going around doing some test pits" to see what the avalanche threat might be. One report posted on a website called PisteHors.com said, "The avalanche risk was 'high' at the time."

But Ashurst knew the mountain, and on the next day, February 27, he studied the snow again. "It had consolidated," a condition that typically lowers avalanche threat. The three men, equipped with avalanche beacons, shovels, probes, and Seigle with his added protections of helmet and Avalung, headed out bearing no sign of being unprepared. "It was a beautiful sunny day," Ashurst remembers of the morning skiing. They skied some runs at Montgenèvre and crossed into Italy, where the Claviere ski resort shares some runs with the French resort.

To get back into France required the men to do a thirty-minute boot hike up a peak, recalls Ashurst. By early afternoon, a little before 1:00 p.m., they had reached "a small prominence" situated on the border of France and Italy. They were standing on their skis at about 7,500 feet, facing north. High enough in the sky, the sun, glancing off the new snow, warmed their backs. The actual name of the run below them was *Cime de la Plane*. From there, by skiing down and to the left, they would return more deeply into the French ski area.

"There had been four or five guided groups ahead of us," Ashurst says, so the men waited at the top, surveying the terrain before them. As a guide, he was always assessing snow conditions. "You ski down and you get to this place where it separates and there are these bowls, many different choices. The bowls to the right had been skied so we decided to ski the one that hadn't been skied." Paused at the top, the men noticed a French gendarme border guard shack near the bottom of the run; they could even spot the occupant, small in the distance like a charming detail in a model train layout.

The day stood before them like a dream. Finally, Ashurst dropped in first to test the snow and give his companions some sense of what the conditions held. He skied no more than two to three hundred feet before stopping to the right of a stand of trees. He looked back up where Seigle and Sawyer stood in the brilliant sun, ready to begin their runs.

Seigle dropped in next and began making turns to the left of the tracks Ashurst had carved. A hundred feet, four or five turns into his run, Seigle saw a crack open in the snow. It was as if the entire mountain face was groaning. For a moment Sawyer saw his friend try to ride the collapsing snow. As it sharpened to a roar and accelerated, the fragmenting snow under Seigle's feet swallowed him.

Above, Sawyer watched in horror as Seigle completely disappeared. The avalanche broke like a massive wave into the stand of trees to Ashurst's left, no sign of Seigle in the white snarl, the man by now trapped under a deep mob of snow, ice, and trees. It was as if explosives had been detonated under the snow and the entire mountainside was vaporized, in its wake a spray of white smoke levitating like a memory in the sunlit air.

As the avalanche soared down toward Ashurst, he reached for a nearby tree to anchor himself. But it was too late. The speeding wall of snow seized him as abruptly as it had Seigle. "The whole bowl came off like a big spoon" had scooped it, Ashurst recalls. "It went instantly from a little trickle of sluff to the whole thing calving off at once." When the avalanche hit him, Ashurst began to plummet down the mountain. With no way to stop himself, he slid nearly six hundred feet, twisting and hitting rock and ice.

"I thought I was going to die," he remembers. He slammed into the trees. The sky above his face became a thick shadow of snow and clouds.

Just below the ridge where Sawyer stood on his skis, he had seen the avalanche's jaws snap about three feet deep, slicing away nearly all the snowpack with it. As the slide gathered speed, he saw it hurtle nearly sixteen hundred feet down the mountain. Sucking up more and more snow, it deepened until, when it came to a stop, it was head high. Before Sawyer, the debris field spread nearly six hundred feet wide. A faint smell hung in the air like freezer burn tinged with pine sap and pulverized earth.

Sawyer traversed on his skis to the right. Loose snow and large slabs filled the side of the mountain where Ashurst and Seigle had been consumed in the slide. One slab had still not released, and Sawyer feared he might set it off. Once he felt he had a clear pathway, he skied down to a flat area and began to make his way back to the left where the avalanche had swamped the trees and he'd seen both Seigle and Ashurst disappear.

Swiftly turning his avalanche beacon to "search," Sawyer tried to find a signal. None immediately registered. Then, as he scouted farther, he spotted Ashurst's backpack, which had been ripped from his body; one of Ashurst's skis, also jettisoned, stood like an arrow in the snow. As he made his way toward the trees, Sawyer thought he heard a voice. Then he heard it again, this time more clearly.

The voice was calling for help. It was Ashurst.

WELL BELOW WHERE SAWYER WAS frantically working to reach Ashurst and Seigle, the border guard occupying the gendarme hut heard and witnessed the avalanche. Seconds later, he was on the phone, summoning a helicopter rescue team attached to Montgenèvre. Stranded on the slope above the gendarme hut, Sawyer and Ashurst almost immediately heard the steady thrash of helicopter rotors as the ship flew toward them above the valley.

As Sawyer sought to dig Ashurst out, the guide said he knew he'd been seriously hurt, but he also knew panic wouldn't help him survive. "My heart wasn't racing," he remembers. "I was cognitive, aware." He recalls a succinct thought: "I'm still on the planet."

Looking at Sawyer, he asked, "Where's Jon?"

"I don't know," he recalls Sawyer's response.

"Well, go find him," Ashurst instructed.

Within minutes the helicopter hovered above, and a small search team descended on ropes from a winch system. Seeing that Ashurst was conscious, they raced to locate Seigle. A few minutes passed, Ashurst recalls, and another man slid down from the helicopter. He identified himself as a doctor. "He says, 'I'm going to dig you out and you're going to be out of here in five minutes.'"

Loaded into a metal basket, Ashurst was drawn up on a cable toward the rotors' whine and downwash of wind. Once inside, he felt the helicopter dip and arc toward the nearby town of Briancon, seven miles away. "From the time of the accident to the top of the hospital was only twenty-five minutes," Ashurst says.

Without the swift work of the gendarme and the rapid response of the rescue team, Ashurst believes the outcome for him would have been far different. He was taken directly into surgery that lasted nearly five hours. His left humerus was seriously fractured and his left brachial artery, which supplies blood to the lower arm and hand, had been severed. A special vascular surgical team had to be called in to transplant a vein from his leg to his arm to restore circulation. Had that not happened, Ashurst's left arm would have essentially died and that would have led to amputation.

Ashurst had broken ribs and surgeons sucked three units of blood from his left lung. His body was bruised, but so was his ski guide's pride in protecting his clients. Once in intensive care, he kept asking hospital personnel about Seigle. "The first two days they told me he was in the hospital in severe condition," Ashurst remembers. "On the third day, I said, 'He's dead isn't he?' And they said 'Yes, he's dead.'"

Sawyer later told Ashurst that Seigle had been pronounced dead at the scene. Despite all Seigle's concerns for safety, and his donning a helmet that day, his impact with a tree had been so grave he was likely instantly killed.

In the hours after Ashurst's accident and surgery, Julie Johnson, who was in the United States at the time, received reports on her husband's condition. "Gary was still in recovery from his surgery," she recalls. "That was one of the hardest parts for me. The phone was ringing every ten minutes. There were so many rumors." But soon the rumors fell aside and the true extent of her husband's life-threatening injuries became evident. "It was the worst phone call you could get," she remembers of the prognosis for his recovery. "He didn't die at that moment. But it was five to ten hours before we would know the outcome of the surgery. . . . I didn't know if I was going to lose my husband."

An accident investigation by French authorities ensued, and the ultimate report noted that at the time of Ashurst's and Seigle's accident, there had been twenty-five avalanches in the region in two days. They did a detailed deposition with Ashurst, who was serving in the professional role of ski guide and was therefore responsible for his clients' safety.

In the immediate aftermath of the avalanche, a team of investigators was sent to the site where they dug into the debris and excavated snow to examine the area of the crown where the slide had been triggered. There they found something completely unexpected. "An unknown fact at the time," Ashurst says, "was that there was a spring under that spot." The tiny stream had been freezing and going through microscopic melts and refreezing beneath the snowpack, destabilizing it into what Ashurst says was "a perpetual trigger." Recalling the sight of Seigle making his first turns from the top that day, Ashurst says Seigle "skied right into a trigger."

In the end, Ashurst was cleared by the French investigative panel. But in the coming weeks, he found himself revisiting his memory of the avalanche and of his friend Seigle. It was a memory that might have led any mountain guide to become locked in deep guilt and depression. "Jon's wife was amazing for releasing me from any guilt," Ashurst says. "When I got back, she said, 'The biggest thing you can do for me is to recover and do what you do. Jon loved you, and loved skiing with you.'"

Ashurst has led clients on thousands of ski runs in some of the toughest, steepest terrain in the world. The accident with Seigle was the only fatality in his near forty-year guiding career. After so much experience in the mountains, he understands the contract, often unwritten in backcountry exploits, between guides and clients. "One thing about being a guide," he says, "once a guide, always a guide. If anything goes wrong in the mountains, I'm going to be scrutinized more than anyone. I'm going to be held accountable."

Yet the most ardent skiers, longtime clients of Ashurst's, returned for years for a week's worth of runs down La Meije. They came season after season, if only to taste the high-altitude air again, to exalt in an afternoon run down seven thousand feet of pitch to the tiny town and its valley and frozen lake below. "Clients would say, 'You're so lucky to be doing this.' I don't think luck had anything to do with it. It was an aspiration I had that I turned into reality."

Sometimes he thinks about the notion some people have of backcountry skiing and the risk of avalanches as a form of cheating death. "There are definitely risks we take out there," Ashurst says of backcountry skiing in places like La Grave, "and there are risks we may not even be aware of."

But he's at peace with the risks. He's thankful for the life he's led in the mountains, the incredible sunrises he's seen as he makes his way to the peaks lit by alpenglow, and the vast white swaths of fresh snow waiting for him and a small group of skiers to draw lines on unblemished slopes in the gold morning light.

Since his accident in 2006, Ashurst has returned to La Grave three times, and he continues to guide some regular repeat clients. Until recently, he guided with a backcountry snowcat operation in Idaho. "My father, when I was in recovery, he was slipping away, really on his deathbed," Ashurst recalls of a day just after his accident. "And he asked me to promise I would give it up. I told him 'Dad, that's not a promise I'm going to make, because I'm not going to keep it.'"

20

BACKCOUNTRY FREERIDERS: BOGEYMAN FEARS, COWGIRL LINES, AND CHUTE DROPPERS

Watching a Teton Gravity Research film of professional snowboarder Jeremy Jones make an elevator drop into an Alaskan "spineology" session—a death-defying dance along a descending cliff edge—you could swiftly reach the conclusion that Jones is either immune to fear or nuts. Or both. But here is his personal philosophy about overcoming danger and fear in the backcountry: "The most important decisions you make in the backcountry are the decisions you make before you even leave the house."

Jones, who lives in Truckee, California (not to be confused with the snowboarder by the same name who lives in Utah and had his legs broken in a January 2017 Uintas avalanche), is the founder of Jones Snowboards, and one of three brothers who cofounded Teton Gravity Research in the late 1990s. He is also one of a very few snowboarders in the world who have combined alpinism with backcountry riding to achieve a sorcerer's level of mastery. Nuts, he is not.

"I think fear is super important," he says of his attitude in the backcountry. "We're going into these places where it's a zero mistake situation." Jones, like professional skiers Angel Collinson and Noah Howell, has taken his sport into the most rarefied air of performance without crossing the "bleeding edge," the line when control evaporates and the rider pays the price of life-changing injury or death. Unlike the nonprofessional practitioners of backcountry riding, a snowboarder like Jones—or skiers like Collinson and Howell—builds his mindset around safety. In so doing, he encourages young riders to both push their own envelopes and carefully curate their fear.

The distinction Jones makes is between "bogeyman fears" and those which may be genuine because they're tied to objective mountain hazards—avalanche signs, bare cliff bands, crevasses. "Bogeyman fears," he says, are those that arise in the predawn darkness when you are crossing a snowfield and the cracking of ice underfoot conjures images of plummeting into an icy prison.

"We would have days in Alaska," Jones recalls, "snow's stable . . . we're going to climb this mountain, weather's calling for sunny skies, and we need to start at three in the morning. . . . And early on we'd get to the bergschrund, which is always scary to walk across because you're basically walking across a crevasse, and we'd be approaching that in the dark. And we'd be hearing the mountains rumbling—in Alaska it's common to be around seracs—which are always making noise. We have these scenarios where we're like, 'I don't know, man, something changed last night, I'm not feeling it.' We'll turn around, we'll get back to camp just as the sun's rising. We'll look at where we were and we're like, 'We were in the perfect spot, it's the perfect day, and we talked ourselves out of it.' That's what I would call 'bogeyman fear.'"

You don't ignore it, Jones concedes, but you don't let it overpower your emotions and earnest decision making. "It's this conversation with fear the whole time. And then as you start moving up the mountain, it's getting steeper and you feel like you're on the edge of the world, but you have crampons and an axe which is super good at keeping you on the mountain. And naturally fear comes in. And it's like, 'Is it bogeyman fear, or did it get too warm and we're starting to see [problematic] snow?'"

IF RELYING ON PAST EXPERIENCE is a part of sifting through the emotional layers of fear, Jones says, it's also ultimately about having crystal-clear judgment that isn't blinded by the sunk cost of an expedition's efforts. "The art of climbing and riding these things is figuring out when to be persistent and grind through hard spots and be really committed to this goal, but having the know-how and ease to realize when the mountains go from green light to red light. And then walk away from a dream line, that you've spent weeks on, with very little thought. . . . And being totally at peace with that."

Jones has learned from his experiences in Alaska, California, Wyoming, the Himalayas, and other ranges that there's a time to react to the danger signs. "The mountains are sometimes screaming at you 'Get out of here!'" There are risks that are always present. "Make no mistake about it, you're going into a wild, uncontrolled place and there's risk involved. Welcome to wilderness. . . . The harsh reality is, and I wish this wasn't the case, accidents happen." He accepts risk but shuns any rider who is clearly reckless. "Those are the ones that drive me nuts, that make me crazy."

Jones's backcountry snowboarding philosophies about fear, risk, and decision making are rooted in his larger life view. He's ingrained his personal values in the business model for Jones Snowboards, where the focus is on innovation, safety, and protecting the environment. "I would say the average snowboard company would spend about eighty percent on marketing and twenty percent on R&D. We spend eighty percent on R&D and twenty percent on marketing."

That direction was planned from the start, says Jones. "I've worked with dozens of [outdoor] companies. When I started my company, I had the idea that I'm going to make this company how I believe in things, and the ethos I believe in. It's either going to work, or it's not. And I was at peace with that." In one instance, in the summer of 2018, the lead photo on Jones Snowboards' website wasn't of a new splitboard design, or a snowboarder shredding a big line in Alaska. It was of Jones, company employees and others, dressed in suits and business attire while lobbying for environmental protections aimed at preserving winter climate conditions. This "actions instead of words" tenet extends to Jones's personal life.

"I value my quality-of-life index way over my financial index," he says. With a travel schedule that keeps him headed to some of the world's wildest mountain terrain, coupled with his company duties, Jones has moved into more of an "advisor level" with Teton Gravity Research, though he still appears in TGR films. "About TGR, the important thing to know is my brothers and I fell in love with the mountains at a very young age." The origins were about "taking our riding as far as we can and doing it safely. TGR, that's what it was born out of and is its heart and soul."

Now forty-three, and in a professional sports world where young, backcountry riders embrace and retire from stardom in their early twenties, Jones still charges hard and shows no sign of easing off on the throttle: "I jokingly say that in pro snowboarder years I'm about three hundred years old."

BACKCOUNTRY PRO RIDERS, LIKE ATHLETES in most sports, tend to have their own lexicon for describing feats of one kind or another. "Cowboying" is an extreme mountain run with seeming disregard for safety, yet doing so with style.

"Cowgirling" might apply to just about any run Utah native Angel Collinson takes. Her skiing has become a staple in TGR films, and hearing the whoops she elicits from her male pro-skier costars, like Sage Cattabriga-Alosa and Ian McIntosh, you have no doubt Collinson's cowgirl lines are as outrageous and polished as any put down by the men. Her skiing transcends athleticism; she's making art on her way down.

Now living in Girdwood, Alaska, Collinson has become accustomed to the far northern maritime weather that sends the thermometer dropping like a fast rappel and smears the sky gray with clouds carrying moisture from the Gulf of Alaska. It can be a long wait before "the milk comes in," as she calls a fresh snowfall, followed by blue skies. But then she climbs aboard a helicopter to go shoot her ski descents in the Chugach and other Alaskan ranges. A two-time champion of the women's Freeskiing World Tour, and two-time winner of *Powder* magazine's "Best Female Performance," Collinson has achieved ultimate status within the international women's professional ski ranks. But her skiing is so dominant, so aggressively muscular and graceful at the same time, performance categorizations as "female" or "women's" seem irrelevant.

Collinson is one of the most elite, elegant ski mountaineers on the planet. Not to mention that she blends her skiing with an intellectual quest to understand how the psychic assets of handling fear can be translated to other life pursuits. "I am driven by an uncontrollable urge to go deeper, to find out what lays inside me, to find out how to listen to my inner voice and the signs presented to me by life," she wrote in an athlete's profile essay for The North Face, one of her sponsors. "For me, it's not about skiing as what I 'do'—it's about why I do it. I do it to push myself. To see where my limits are. To be intimate with fear and not let it rule my life."

In mapping her journey as a skier, Collinson has broken down her phases, relating her ski mastery to Malcolm Gladwell's *Tipping Point* proposition that ten thousand hours as a musician, painter, or writer—or skier—is the ante to that mastery. To get to her level, there's a sort of physical leap from the ten-thousand-hour platform into a weightless realm where the orienting agility of a cat merges with a gyroscopic precision that's attainable by only a very few.

"Many people get a rush from overcoming fear and that is their inspiration to follow madness," Collinson wrote in *The North Face Journal*. "I personally don't enjoy taking risks. I like living in the comfort zone. I want to feel capable and competent, and standing on top of a really challenging ski line makes me feel anything but that. And that is why I do it. I don't think we ever grow if we stay in our comfort zone." The emotional tension Collinson wrestles with in skiing is something she seems to have been born into. Her father, Jimmy, worked at Utah's Snowbird ski resort as a ski patroller and was one of their most seasoned avalanche control people. The family lived in employee housing, her bedroom a five-by-twelve-foot "closet" she shared with her brother, pro skier Johnny Collinson. The kids were homeschooled, partly so they would have the liberty to accompany their parents, in a van, on lengthy road trips to rock climb all over the American West. "Van life wasn't cool back then," she says. "It may be now. But back then I was the weird kid. . . . I loved being outside though."

If Collinson was baptized early as a skier, a climber, and all-around mountain girl, there was a parallel taking shape; she defined herself as a "student" of the world, particularly the outdoors and the environment, but also in terms of literature, music, and other art forms. "My profession is in the mountains, but my hobbies are not." To that end, she's always looking for opportunities in Alaska to attend symphony performances, art shows, poetry readings anything that can help expand her consciousness as she charts her future. Her internal dialectic with fear isn't unlike that of Jeremy Jones when he makes distinctions between "bogeyman fears" and those which emanate from genuine empirical reasons for fear. "Fear shows up in two different ways. . . . A healthy fear like the fear you feel before you take a test . . . nervousness in the belly and throat. . . . Then there's a fear that's like an intuition. And that's the 'Don't do this!' fear." It's that latter fear Collinson explores to gauge where it's originating and if it's something she needs to heed or just box up in a corner of her mind.

ALWAYS PRESENT IN HER MENTAL dialogue, even if she's gotten past the days of her worst memories, is the 2011 death of Collinson's boyfriend, Ryan Hawks. A competitor on the men's Freeskiing World Tour, Hawks was on a competition run in Kirkwood, California, chasing a particularly difficult line. He catapulted off a huge cliff and went into a backflip, aiming to nail it for a spot in the final. But he missed his landing, suffered extensive injuries, and had to be evacuated by helicopter to a Reno hospital.

Collinson had to compete the next day in the women's tour events. So flawless was her skiing, she took home the winner's trophy. Afterward, she rushed to the hospital and sat at Hawks's side until he passed away the next morning.

"What happened with Ryan was not a mistake because he did those kinds of things every day in his skiing," Collinson told ESPN in an April 2011 interview. "He landed on a rock, and sometimes it happens, and sometimes I really think it's your time to go. If you know your limits and you're always aware of them, then you shouldn't be too scared. Because there's always a danger of something happening."

While Collinson went through a bout of deep grief over Hawks's death, she ultimately found motivation in her memories of him. "On days when I'm sad and sitting on the couch drinking tea," she told ESPN, "I just think, what would he say if he saw me sitting around feeling sorry for myself, crying and watching it snow? He would think it was the most ridiculous thing ever, and he'd be like, 'Angel, what are you doing?' So sometimes, for him alone, I'll get off the couch and go out and have a great day."

One concern she has regarding risk is that a lot more young people want to ski and snowboard in the backcountry, which brings exposure to more marginal conditions for themselves and surrounding riders. "Backcountry skiing is definitely becoming a cooler thing. I would also say lift ticket prices have gone way up" and that is driving some of the trend. More traffic in the backcountry is one dynamic Collinson sees, but her misgivings derive from a sense that some people aren't prepared for it. "Just as concerning to me [as more people] is the level of skiing some people are taking into the backcountry. . . . It takes a lot of time to become a good skier. . . . A lot of skiers don't, I think, have the ability to ski out of an avalanche." Her own experience with avalanches has been limited. "Knock on wood, I haven't really had too many experiences with avalanches. But I'm always in avalanche terrain. It's a constant," as is the need, she adds, for "snowpack assessment."

When Collinson eases up on her personal quests, it's her time in the mountains filming with TGR, and her friendships with other elite skiers, that she finds most rewarding. "Every time you go on these trips, you come away with the fact that the most important thing is the people you're with. Flat out. Every time."

AFTER SITTING IN AN REI store watching world-class mountaineer Andrew McLean give a presentation on his just released Utah ski guide *The Chuting Gallery,* Noah Howell's first reaction to the serious exposure to danger on McLean's recorded runs was: "Why would anyone ski this stuff?" Not long after that, though, Howell veered 180 degrees and decided *he* would be one of the first skiers in Utah to answer that question. And then he skied every descent in the book.

A tall, powerfully built man with glasses, trim sideburns, and neat, outdoorsy clothes, Howell conveys the look of a young college professor teaching recreational management more than that of a pro backcountry skier. He's deceptive in both his looks and the modest, humble demeanor he projects when talking about his ski achievements. Unlike Jeremy Jones and Angel Collinson, Howell's name isn't a marquee one; he appeared in some ski films that he made with his brother in the mid-2000s but their showings were underground cult favorites rather than mainstream productions with splashy openings in mountain town movie theaters.

Howell occupies that sometimes dubious but nonetheless praiseworthy distinction of being a backcountry skier's backcountry skier. Should there be any doubt about his bona fides, however, in 2013 he was named by *Backcountry*—the bible of the sport—as one of backcountry skiing's "Icons Among Us: 50 Living Legends." Howell is that rarest of pros—the soul skier or surfer who doesn't need to compete to attract a devoted following. Take, for example, his descents in Alaska, when Howell and his partner, Adam Fabrikant, were the only ones who climbed a particular mountain in the *entire season*. He's the athlete whose reputation coasts along nicely on ardent rumors about his expeditions, chance encounters on the mountain, and glimpses of his majestic lines.

Howell is that mountain lion in the shadows, traveling lightly, seeing you perfectly when you don't even know he's lurking. When I met him in a Salt Lake City coffee bistro, I had arrived early and had no real idea what he looked like. Scanning the faces before me, I tried to pick out the guy who looked like what I expected. Just beyond the coffee bar, I'd noticed the man but didn't give him a second glance. He didn't look like he'd rolled out of a sleeping bag with a bad case of bedhead.

Then I heard someone shout: "Howell!" He removed his ball cap to reveal a shaved head except for a stubble of hair down the middle like a tonsured mohawk. This I took to be evidence of his standing as a backcountry badass. That and the snake tattoo like a bracelet wrapping his left wrist. We sat and talked about his spring trip to Alaska with Fabrikant, a Jackson Hole skier who works in the summer for Exum Mountain Guides. Howell insists Fabrikant is notching ski mountaineering feats that can only be compared to the most outrageous lines in Chamonix, France.

The pair traveled to Alaska in May 2018 without a specific mountain in their crosshairs. "Our goal was to ski something steep and fun," Howell says. "A little bit 'out there.' Most people, however, wouldn't consider it fun." Among their potential targets were peaks in Alaska's Wrangels, Revelations, and the Tordrillo Range. They had skied Denali in 2005 and were interested in Mount Foraker, but flying in they could see through the plane's windows that the mountain was sheathed in ice. After some debate, they chose Mount Hunter, a 14,573-foot peak in Denali National Park, eight miles south of Denali itself. It's the third highest peak in Alaska, sometimes referred to by its native appellation, Begguya, which means "child of Denali," though nothing about it is juvenile. An online description of the summit notes: "Long corniced ridges extend in various directions; between them are exceptionally steep faces."

Howell and Fabrikant started their approach to Mount Hunter at about six thousand feet, the bottom of the Ramen Couloir. The path initially carried them through icefall terrain. At about eight thousand feet they set up camp on an open glacier. When they were ready to go for the summit, they left at about 11:00 p.m. because the Alaskan summer nights are just "darkish," says Howell. Even at that hour there is reasonably good light and visibility if the weather is clear. The first pitch involved a three-thousand-foot hike, at the end of which they began to boot up the steeper, remaining distance to gain the 14,500-foot summit's west ridge.

"We had an incredible day, no winds, no clouds," Howell recalls. But the placid conditions on top emphasized the need to not become complacent. "When you fall down that mountain—nine out of ten directions you're pretty much dead." Because of the foregone conclusion in the event of falling, and in the interest of keeping down weight, they did not even carry avalanche beacons, shovels, or probes.

Reaching the summit about 9:00 a.m., Howell and Fabrikant "hung out" for an hour, taking in the morning's show of snow-clad peaks. After a ten-hour hike and climb at altitude, it was a chance to catch their breath and shift their mental focus to the ski down. Then it was time to click into their bindings and go. "We skied down the summit ridge and did two rappels to get down some ice," Howell recalls. After that, they felt like the mountain opened its palm to them. "The conditions were amazing, corn snow," he says. "Usually you're not expecting great ski conditions, but this was really fun." "Fun" meant an uninterrupted ski of nearly 6,500 feet to base camp. So many turns, most skiers—even experts—would find their knees turning numb.

HOWELL HAS MADE THE TREK to Alaska every year for more than a dozen years. But it isn't every time he experiences a descent like the one on Mount Hunter. Too often, he and his crew have been denied a mountain by stormy weather, sent home without coming close to accomplishing what they had planned. But sometimes, even when a trip's preparations include the likelihood of bad weather, Howell has been rewarded. In 2015 he and some friends skied 12,218-foot Mount Cook, the highest peak in New Zealand, a mountain infamous for bad weather. But other conditions met them on arrival. "We showed up and just teed off," Howell recalls. After Mount Cook they ticked off more peaks, so much so that when they encountered skiers from Down Under, they were greeted with: "Oh, you are those Americans who are just skiing everything."

Howell shares the concerns of Jeremy Jones and Angel Collinson: the effects of climate change on mountains, increasing human traffic, and the resulting dangers as these dynamics collide. "It's very different than when I started," he says. "We're getting less snow in the valley, more rain. It's pushing people up to higher levels. It's forced everybody up into the same areas." There's another aspect that may lead to more dangerous situations. With the snowpack undergoing climate stress not historically experienced at higher elevations, there is also the presence of slope angle as a rider goes higher. Combine snowpack instability with an imperative to hike to higher terrain and it's an accelerated equation for avalanches.

At forty-two, Howell has entered midcareer as a pro skier. Once a competitor on the national telemark freeskiing tour, he is no longer willing to travel if it doesn't involve an especially challenging ski-mountaineering destination like Mount Hunter with its alluring blend of alpine-style climbing and skiing. He makes a living through backcountry guiding, teaching backcountry skiing courses, editing ski videos, and writing ski articles for magazines like *Backcountry* and *Ascent.* Occasionally, he puts on slideshows about his trips. Mountaineering company sponsorships from Black Diamond, Scarpa, Julbo, and others help fill in the gaps.

"I'm surprised I've stayed in it this long," he says. "But it doesn't get old. The guiding, the writing, the sharing with other people." With experience and maturity, Howell has developed a more objective filter for fear. "You have to spend a lot of time approaching fear to understand . . . [when] it's just an emotion that doesn't have any backing. I've had times when I'm just scared and don't know where it's coming from." Staying out of danger zones, and having the knowledge to spot them ahead of time, is a key part of backcountry judgment for Howell. It's one reason he's managed to largely avoid the avalanche dragon and never suffer the most serious consequences.

"I've triggered a lot, and skied out of them. I've never been buried. I've been tumbled and spit out. You can't understand the power—and powerlessness—you are experiencing until you've been in one." He breaks a smile. "It's a fun game. It's amazing."

21

A TASTE OF SNOW ON THE WIND

One of the truths avalanche forecaster Craig Gordon must reluctantly accept about his job is that he can warn people that avalanches lurk in the backcountry, but he usually can't pinpoint where they are hiding. Detecting the signposts of an avalanche is a proven science that Gordon's onetime boss, Bruce Tremper, and others have written about extensively. All you need is the mind of an actuary, the eyesight of a geologist, and the probability mindset of a Vegas gambler. Then there's Gordon's 1 percent of unknowability. Call them the outlier snowflakes: you can spend a lifetime reading snowpack and, even with such experience, coupled with a keen intuition, you can miss the claws lurking just beneath the surface.

Every fall since 2007, Gordon convenes a group—forecasters, scientists, meteorologists, snow professionals, ardent backcountry skiers and boarders—and holds the equivalent of a two-day mass snow séance, although some might feel it's more akin to a tent revival for backcountry supplicants. It's an effort to tap into the otherworldly dragon harbingers they all know will soon be ghosting through the coming winter's snow. Look over the crowd and you might consider it a sort of cultish conclave for people inclined to wear fleece.

For the winter of 2017–18, Gordon held his Utah Snow and Avalanche Workshop on the first weekend in November at Snowbird ski resort. When he initially began the workshops, they were attended by a handful of hardcore backcountry aficionados who were more like extended family from around the Salt Lake Valley. To Gordon, it felt clubby. But the November session drew a somewhat larger crowd, reflecting the growing interest in the backcountry. Gordon watched the room fill to nearly a thousand people.

Anyone familiar with a weekend professional conference knows about agendas, breakout sessions, video presentations, and trend discussions. Gordon's conference agendas have a certain cacophonous variety to them—law enforcement meets high-wire walkers meets snow geeks. The sessions had titles like "Explosives Handling" presented by an Alcohol, Tobacco and Firearms expert; "The Changing Face of Backcountry Riding" from a guy with a group called Alpine Assassins; "Looking at Snow Patterns Like a Pro" from a PhD university research scientist; "Where's Your Partner?" about an "alarming number of 'solo' avalanche deaths"; and even a nod to the fractured state of American politics: "Flake News: Making the Ski Industry Great Again in an Era of Negative Media."

The problem Gordon and other snow professionals wrestle with each year, as they prepare for winter, isn't actually avalanches—it's the people who will run afoul of avalanches. The experts' annual coming together is a little like evangelists updating and fine-tuning the gospel with the overarching theme that prevention is always better than rescue. Gordon and others are constantly laying out and expanding schedules of avalanche awareness sessions. He'll meet with snowmobilers at a Polaris dealer or a group of elite skiers gathered in a conference room at a ski design company. The more Gordon can disseminate, in the fall and early winter, his latest guidance on avalanches in the Wasatch mountains, the better the chances people will venture into the backcountry at least wary of avalanches, if not scared out of their wits by his stories.

As a practical matter, few backcountry skiers and boarders tend to be timid souls, no matter how tragic the stories of avalanche deaths. It's just not true to type. But at least Gordon can bring visuals and technical details to his awareness sessions that will instill deeper knowledge into backcountry devotees. After education, and the early winter public awareness campaigns, such as Gordon's TV appearances, the snow professionals focus on the avalanches. It's not a question of *if* the winter will bring snow slides; it's a question of how many, in what places, how extreme they'll be—and how many people will get hurt or killed in them.

Avalanche deaths in the United States, a statistic Gordon and his colleagues regularly review at gatherings such as his annual workshop, are recorded each year by the Colorado Avalanche Information Center (CAIC) and reported on their website, Avalanche.org. From the winter of 1998–99, when the CAIC began recording avalanche deaths, through 2019, deaths have ranged from a low of eleven in the winter of 2014–15, to a high of thirty-six in the winters of 2007–2008, and 2009–10. The four deadliest winters were in 2001–2002 (thirty-five), 2007–2008 (thirty-six), 2009–10 (thirty-six), and 2013–14 (thirty-five). Two winters recorded an unusually low number of fatalities—eleven in 2014–15, and twelve in 2016–17.

Beyond the six years of extremes reported by the CAIC, the remaining winters have shown remarkably consistent numbers. Deaths in those years range from twenty-two to thirty-four, with the average being just under twenty-seven. For the years through the winter of 2017–18, the average is exactly twenty-seven. In the heavy winter snowfall of 2018–19, twenty-five people were killed, just barely under the previous years' average.

One thing that stands out for experts is the relative consistency of fatalities, even though backcountry skier and boarder travel has grown almost exponentially. Because there are no requirements for backcountry users to "check in" or report their movements on trailheads and skin tracks, numbers are general estimates at best and are drawn from crowded trailhead parking lots and other sightings in the backcountry. Sales of backcountry gear tracked by manufacturers and outfitters have exploded. In the winter of 2017–18, more than twenty-four million dollars of alpine touring gear was sold, 30 percent more than the previous winter, according to snow sports industry sources.

While deaths in the backcountry haven't sharply increased, it's not known exactly how many avalanche accidents go unreported. Websites like the one maintained by the Utah Avalanche Center carry many accident reports that come from ski resorts and from backcountry skiers who are keen on warning about slide-prone slopes. But it's almost a given that many narrow escapes from avalanches go unreported by skiers and boarders who don't want to take the time to file a report or may even feel sheepish about admitting to such an encounter.

Gordon approaches a new winter's inevitable snow, and the infinite formations it takes, as if it's something coming alive. He knows it's breathing down under there. Every day, its crystalline structure is changing as if to renew its icy skin, or it's beginning to suffer the onset of disease. Gordon knows that sometimes its impulse is to explode through the surface like lava boiling up from a volcano, building up pressure, ready to snarl as the mountain shrugs its malevolent shoulders. He's seen what can happen when there's an evil chemistry of snow, wind, sunlight, and gravity.

ALL THIS PREPARATION FOR WINTER requires a certain resolve as the first snowflakes begin to fall. On the one hand, Gordon is a fanatic skier at heart, and seeing the inaugural snowfall of the season gets him excited about first tracks in the backcountry; the high school kid in him, who first saw blue-jacketed avalanche patrollers at Alta, awakens and still doesn't regret skipping college classes to ski. On the other hand, he has to plug into the machinations of assembling daily advisories.

Gordon has to make sure his mountain instruments—SNOTEL depth reporters, wind and temperature sensors—are working fluidly. They're like mechanical trolls that live in the rocks and report data: they need to be woken up and readied for the winter's coming misadventures. If the summer has made them temperamental, Gordon may get false information that will make his forecasts unreliable. He has to summon his network of snow spies out of summer hibernation—the backcountry skiers and boarders who send him photos, videos, and on-site reports that are essential to his reports. He has to prepare his own backcountry gear for frequent patrols into the backcountry, where he'll dig snow pits, eyeing and prodding the snowflakes, prospecting for the quarrel between old, rotting snowstorms and new ones.

Assembling daily avalanche advisories is a bit like making home movies for YouTube or Vimeo. They're meant to be consumed for a limited time, appreciated for their perishable value. But when big avalanches happen, the production becomes more like an IMAX movie. Everything gets bigger—the number of people involved, the vehicles, air traffic control of helicopters and fixed-wing aircraft, cameras, snowmobiles, even police cars and fire trucks with sirens blazing.

In an avalanche prone area like Little Cottonwood Canyon, the Utah Department of Transportation (UDOT) musters every weapon it has to combat avalanches as it had to in early February of 2020. A huge snowstorm caused the complete shutdown of the canyon for two days; canyon residents and guests at hotels were ordered to go on "interlodge," the town of Alta's polite way of saying "lockdown." While no one was killed or seriously hurt, avalanches measuring up to fifteen feet deep swept down the mountainside, burying the main road and enveloping cars in the Alta ski area parking lot.

All this infrastructure for responding to an avalanche is practically unknown in the eastern United States; it only comes to the public's attention when there are multiple deaths or other particularly unusual circumstances. In the East, hurricanes and floods figure as major weather events. The notion of an avalanche depositing a fifteen to twenty-foot wall of snow on a highway, shutting it down for days, is not a common hindrance to the morning commute in Delaware or Connecticut. But in a mountainous state like Utah, it's not only a reality, it's an event for which the state itself prepares and commits serious resources to each season.

Liam Fitzgerald served for sixteen years as the head of the UDOT crew that oversees avalanche mitigation in Little Cottonwood Canyon. At Gordon's 2017 workshop, he gave a presentation about some of his experiences in the canyon. "It has the highest avalanche danger scale rating of any road in the United States, and perhaps North America," Fitzgerald says of the road that skirts Snowbird and Alta ski areas.

To illustrate the threat level of avalanches in Little Cottonwood Canyon, as compared to surrounding locations, Fitzgerald notes that in a given winter howitzer shells will be fired into the Little Cottonwood mountain snowpack about five hundred times; in nearby Big Cottonwood Canyon it's about forty times a season, and in the Provo Canyon area about fifty times. A twenty-seven-year veteran Snowbird ski patroller and avalanche mitigation expert for that resort, Fitzgerald has seen it all. His presentation was titled "Advice to My Younger Self: A Personal Reflection on the Role Mentors Played over My 48 Years as an Avalanche Professional."

Looking back, Fitzgerald marvels at how many mistakes he made as he learned the techniques of avalanche mitigation. Some of his education came from older experts, some came from just being on the snow, and other lessons came from close calls when he was caught in avalanches himself. "I've been caught in a number of small avalanches and one or two that I would never expect to be that lucky to survive again," he says. "I've made some bad mistakes and nearly paid for it with my life."

In his UDOT work, Fitzgerald was like an air traffic controller for the road threading Little Cottonwood Canyon. Keeping the traffic moving was the goal, but blizzards and snow slides frequently played havoc with motorists. "Closures can last anywhere from an hour to several days, though several days is unusual. . . . It's a twenty-four-hour-a-day problem." With a team of three full-time avalanche mitigation people for Little Cottonwood Canyon, Fitzgerald relied on military artillery, meteorology reports, computer programs, and word of mouth. "In the last five or six years, there was a movement of getting away from military armament," he says of the howitzer firings. That movement was supported by an Infrasonic Avalanche Detection System, a network of three towers and an array of sensors installed on the slopes around midcanyon where avalanche slide paths were particularly active. The sensors are able to detect low-frequency sounds attributable to avalanche release. Coupled with software that can "read" those sounds, Fitzgerald says the system provides data on the magnitude of a slide and how far it has run. It can even send an alert and data to the crew's cell phones.

"It's sort of like having a person on your crew with superpowers," he says. But even with superpowers, avalanche mitigation crews are sometimes at the mercy of nature's white wrath. It's why, when he looks back, Fitzgerald emphasizes the importance of lessons passed on from mentors. "I made a lot of mistakes and saw a lot of things go wrong," he says of the inevitable learning curve when you are battling nature. He speaks of a kind of continuum for snow professionals. "There was a wealth of knowledge they transferred to my custody in a fairly short period of time," he says of his early days around avalanche experts. After nearly five decades of such work, he is the one now passing along lessons. "Even if they're only going to use it once or twice in a career, hopefully it prevents someone else" from making similar mistakes.

Sometimes being an experienced avalanche mitigator and mentor to younger professionals means stepping into the line of fire himself. "I was much more willing to put my safety on the line," Fitzgerald says looking back, "rather than that of the people I was supposed to be protecting."

IF THE HISTORY OF AVALANCHE studies and awareness began with an empirical focus on nature (interrogations of the snowpack) and progressed in later years to the "human factor" (faulty decision making attributable to psychological traps), its emerging model is an inquiry into why humans, knowing everything they do from previous avalanche awareness models, still make bad decisions in the backcountry. Call it an exploration into formulating the "hubris quotient" of backcountry riders. Or, put in less scientific language, a study of the activity of arrogance.

Two chief US–based proponents of this new edge of inquiry are Dr. Jerry Johnson and Dr. Jordy Hendrikx. For Gordon's workshop, Johnson presented an overview of the men's collaborative work with Dr. Andrea Mannberg, a Norwegian behavioral economist who works with researchers in a number of countries. The overall team calls their explorations "The White Heat Project." In an interview, Hendrikx, who heads up the Snow and Avalanche Laboratory at Bozeman's Montana State University, said he and Johnson have enlisted in their study about a thousand people around the world. The focus is to try to understand why, once riders enter the backcountry, they make the movements they do and what decisions they make based on terrain, personality type, and group dynamics.

Equipping backcountry riders with smartphone GPS software, Hendrikx and Johnson have produced several thousand recorded ski and snowboard tracks in mountainous terrain. Prior to recording their tracks, riders fill out preseason questionnaires specifying their gender, age, riding level, and behavioral traits. Once a track has been recorded, the riders receive a post-trip questionnaire about their companions' demographics, decision-making routine, and the nature of the terrain they chose to travel in.

When all this information is overlaid, Hendrikx and Johnson can see the topography map in which the riders traveled and spot how the steepness and potential danger of it correlates with the riders' personality snapshots and reported communication style. There are questions in the post-backcountry survey like "Did your group have a clear leader?" and whether "there were frequent terrain selection discussions" and if "the group was willing to make changes to the choice of route/objective." What Hendrikx is attempting to do in overlaying all this information is produce a "geographic expression of risk" and seek to understand how that squares with a given rider's psychographics, particularly his or her professed risk appetite.

Through the accumulation of data on a rider's personal attributes, the study provides insights, Hendrikx says, on "more about who they are, not just where they went." At the heart of this idea is a view, often quoted in avalanche courses, that "when the snowpack is your problem, terrain is your solution," meaning troublesome snow should usher you to lower, less steep terrain. The submitted GPS tracks by participants provide insight into whether this is actually happening.

Hendrikx and the team are looking at backcountry dynamics through the lens of behavioral economics. He reports: "The behavioral economic view of the world is that everything has a cost and a reward, trade-offs and benefits. . . . Typically, behavioral economists look at this in terms of monetary gain or so forth. The other thing we're looking at is social gain, social status, and positionality." The term "positionality" means that as people strive to be more successful in their social environment (i.e., more admired skiers), they raise the level of what is "good enough" for some people but reduce the status of those lagging behind.

Hendrikx says it's these kinds of group dynamics that can govern decisions on whether to ski more dangerous slopes versus those that are clearly safer. With the personal profiles, their research team is trying to gauge the "risk plasticity" that an individual has. One area the questionnaire explores is "how tolerant are you to increasing your risk exposure beyond your most preferred choice." This is "a measure or index of how group pressures might influence decisions beyond what we would choose to do ourselves if we were controlling all the decisions." This kind of breakdown in group communications—and the social isolation of an individual or multiple riders in a larger group—can lead to accidents. The cockpit of an airliner is an example, Hendrikx says, of where the same kind of communication dynamic can develop and lead to accidents. Airline pilots therefore have extensive checklists that raise necessary questions and create back-and-forth evaluations among pilots.

But in skiing that may not translate as readily. "The last thing we want to do when we go backcountry skiing is say, 'Alright, let's pull out checklist number seven and go through the twenty-eight points that I need to tick off before I go skiing.'" Becoming self-aware of risk tolerance, says Hendrikx, isn't something typically talked about in avalanche awareness classes. "We don't talk at all about peoples' risk tolerance and risk appetites. What we're seeing is that a few simple questions in our survey are really insightful in terms of the types of people and types of settings where people are more willing to tolerate more risk."

The White Heat Project isn't just an abstract laboratory exercise for Hendrikx. As much as anything, it's tied to a phone call on a day Hendrikx can't forget: January 5, 2015. Standing on a stage under a large projection photo of a smiling freckled young woman in a knitted ski cap, Hendrikx gave a TEDx presentation in Bozeman, Montana. "It was the sort of phone call," he told the audience of several hundred, "you really never want to get. The phone call informed me that, unfortunately, one of my snow science students, Olivia Buchanan, a twenty-three-year-old undergraduate, had been killed in an avalanche in southern Colorado."

Hendrikx knew Buchanan well. "She made sure of it. She came into my office, probably on the first day at MSU [Montana State University], and told me who she was, what she was doing, why she was there, and where she wanted to go." When Hendrikx and his colleagues work to push the White Heat Project into new areas of examination, he thinks of Buchanan. "It was her death, and unfortunately those of others after," Hendrikx told his audience, "that really motivated me to think about this problem a little bit more broadly. . . . So her death was really a big turning point in my personal research career from solely focusing on the snow science, and really thinking about how we can better understand the human side."

IF THE WHITE HEAT PROJECT is a high-level strategic approach to going to war with "the avalanche problem," as Hendrikx and others call it, search-and-rescue expert Nancy Bockino is like one of the foot soldiers focused on a specific target—communication in the backcountry. Bockino, who also presented at Gordon's avalanche awareness workshop, thinks of the phrase "the human factor." Rather than interpret it as a negative, she advocates for it to be the positive impetus for more questioning, sharing of information, and debate in the backcountry. She asks, who has a better ability to communicate than humans? So why not use it? To that point, her presentation was titled "Embracing Being Human: Freedom in Communication and a Mountain Ethic: Using the 'Human Factor' To Keep Us Alive in Avalanche Terrain."

Bockino's résumé reads like an extensive topo map of all the outdoor disciplines one can imagine acquiring in the mountains. Based in Jackson Hole, Wyoming, she is a veteran of Teton County Search and Rescue, a member of Exum Mountain Guides, a trainer for the American Institute for Avalanche Research and Training, and a member of the American Avalanche Association; she's also the winter ops manager for the Jackson Hole Outdoor Leadership Institute. In the summer she works as an ecologist in the Greater Yellowstone Ecosystem, which includes Grand Teton National Park, one of the country's public lands crown jewels.

Bockino summarizes her view of augmented backcountry expression as "communication passion paired with responsibility." Unlike many approaches to avalanche safety, most of which have arisen from scientific investigation, her approach grew organically from working in search-and-rescue, guiding in the mountains, and through avalanche training. "I really believe people need to be empowered and treated with respect. I like the idea of being positive instead of giving people this list of what they shouldn't do."

Preparing to teach an avalanche safety class, Bockino recalls studying a list of human factors assembled as warning signs for backcountry riders. When she realized her own reaction to the list amounted to a disincentive for traveling in the backcountry, she imagined her students recoiling from it. "You present the list of human factors and it's not engaging, it's not relevant, it's not useful. It's not going to keep someone safe out there." Instead, Bockino believes, people learning skills for the backcountry should be given "a toolbox of choices of how they're going to communicate."

When teaching avalanche classes, Bockino makes a point of urging people out of complacency to embrace leadership roles, even if it means pulling them out of their comfort zone. She assigns students roles such as keeping track of which students have not spoken up or asked any questions. When the student given this task eventually indicates who those nonparticipating students are, it's a chance for everyone to vocalize a point of view. In this way, Bockino says, she seeks to create shared learning and a team mentality. "I try to get people to think about what kind of leadership style they have . . . pushing people towards making sure they realize that when you take on this thing you want to go do in the backcountry, it comes with responsibility to be a leader at some level."

Bockino reveals her own vulnerability by telling stories of mistakes she's made in the backcountry. "Anytime you're honest with people, they can learn from you. They feel safe around you knowing that you messed up and that you're willing to be open about it." Mainly, she encourages people to understand that when they go into the backcountry, they need to carry with them—just like they carry a beacon, shovel, and probe—a heightened sense of observation, questioning, and willingness to challenge each other. Their communication level needs to be as important as their riding level.

"We all fall short," Bockino says, "but how can we do it better? Expert-level communication is a challenge in all parts of our lives."

22

OCTOBER MORNING ON IMP PEAK, MONTANA

The town of Bozeman, Montana, with its population of forty-five thousand, has been known for decades to have a disproportionate share of world-class skiers, mountaineers, rock climbers, and assorted outdoors fanatics. Among them are those accomplished on the world stage and others still waiting in the wings for global acclaim. Bozeman is the perfect breeding ground for this sort of aspirational mastery of nature's funhouse. The town is surrounded by the Gallatin Range, and one would be hard-pressed to ski and climb all its peaks in a lifetime, much less during a dedicated apprenticeship. Then there is the Montana spirit toward individualism, if not outright idiosyncrasy. If your spirit is as wild as the topography, you should be allowed to map your life's pathway, no matter how rebellious it might be.

Hayden Kennedy, twenty-five years old at the time and already a world-class mountaineer, met Bozeman native Inge Perkins at an ice-climbing festival. Tough, willful, adventurous, and generous with her spirit and friendship, Perkins was an accomplished climber with sponsorships from Mystery Ranch, a mountaineering backpack company, and Scarpa, a top maker of rock-climbing shoes. As a backcountry skier, she had also made her mark, skiing the Grand Teton and winning ski competition awards. At one point she renewed her commitment to the out-of-bounds, saying, "I have fallen back in love with backcountry skiing and ski mountaineering." Perkins stayed faithful to her mountain pursuits through a nomadic lifestyle. Despite deep connections to Bozeman, a town she returned to often, she once described her place of residence as "where my truck takes me." She was, as a friend put it, a "blithe, angelic badass."

Kennedy, disarmingly laid-back, shared a comrade's wanderlust with Perkins, only his travels took him around the globe to big mountain, alpine-style bivouacs and multiday climbs of sometimes life-threatening magnitude. More poet than braggart, he once said of climbing with partners: "When climbing partners connect so well in the mountains, they become like musicians: there are no words spoken, just the acts of several people communicating through riffs; the sharp hit on the snare drum, the pulse of the bass; the single sound moving forward." Kennedy had been named an ambassador for Black Diamond, worldwide maker of technical climbing gear. Some in the global climbing community viewed him as almost a prodigy. In July 2014 the magazine *Elevation Outdoors* published a story about Kennedy proclaiming he might be "the best young climber on the planet."

Hayden and Inge fell in love immediately. Doug Chabot, a friend of Kennedy's and director of Bozeman's avalanche forecast center, characterized the couple's romance as one like Romeo and Juliet. So quickly and deeply did they became mountain soulmates, the two decided in 2017 to take a year together to travel and ski and climb in Europe and all over the American West. Perkins had suspended her studies at Bozeman's Montana State University so she could be with Kennedy. Kennedy, who had been on a publicly low-key but tenacious campaign putting up world-class climbs in South America, Pakistan, and Mexico, among other destinations, slipped out of mountaineering's limelight to be with Perkins.

Midsummer, the pair returned to Bozeman. Perkins planned to resume her studies toward her math degree and prepare for teacher training in Belgrade, Montana. Kennedy was to begin studying for an emergency medical technician certification. The young couple found themselves in Bozeman back among family and friends. Perkins's father, Steve, was a PhD professor in MSU's civil engineering department and her stepbrother, Christopher Carter, a filmmaker and city planner. After all his extended travels, Kennedy, an only child, was just a twelve-hour drive from his childhood home and his parents, Julie and Michael Kennedy, in Carbondale, Colorado. He had repeatedly returned to Carbondale, where he had a wide circle of friends, many of them former classmates at the Colorado Rocky Mountain School.

With Inge's birthday approaching, Kennedy decided to throw her a surprise party and enlisted their close friend, Kelsey Sather, to help out. She and Kennedy sent invitations to friends in Colorado, Oregon, Wyoming, and other places in Montana—an intended gathering of Perkins's extended tribe. They even started planning special birthday cakes for the party, to be held in Lander, Wyoming, where Perkins had visited many times and had a second orbit of friends. Kennedy kept the party planning secret; so quiet had he been about it, the day before the party Perkins baked a cake for what she assumed would be an intimate, low-key celebration with a handful of guests.

The night before the surprise, Perkins, ever ambitious and pushing herself to new levels of athletic prowess, announced she was going to do a near eighteen-mile traverse in Wyoming's Wind River Range. "We both know Inge's an incredible athlete, don't get me wrong," Sather recalls Kennedy telling her over the phone. "But that traverse is a full-day endeavor. She'd be lucky to be back by nine or ten."

Worried Perkins would miss her own party, they got a friend to lure her to a nearby day climb and the party was on, Kennedy reported to Sather. An ardent baker, Kennedy set about mixing up two birthday cakes.

"It would have been a little strange if you didn't show up to your own party," Sather recalls telling Perkins, referring to the small gathering Perkins was expecting. Perkins agreed, saying she figured people would still have fun. It was a classic Perkins sentiment. While her friendship net had been cast wide over the years, as she "pranced around the world," her own description of her travels to climb and ski, Perkins called her solo life moments as living in her "hermit world." She had once sent Sather a Christmas card on which she'd written: "This might sound kind of funny, but one thing I truly admire about you is your awkwardness in social interactions sometimes . . . seeing that you can . . . build meaningful relationships while not socializing perfectly has really helped me with my insecurities."

But if Perkins sometimes had doubts about her personal relationships, the surprise birthday party success must have refuted her misgivings. Sather remembers Kennedy grilling sausages and burgers for the guests—climbing and skiing friends, older adults, kids. "They had built a tribe together," Sather says. "They were loved, wholly and widely."

Over nearly a decade of friendship, Sather and Perkins had grown close. Sather did not know Kennedy quite as well, but she was discovering his endearing traits—his self-effacing expressions, his probing at life's deeper questions, and even his occasional goofiness. "There was one thing I did know: he *loved* her," Sather says. "He loved her with a care that had him texting and emailing me *daily* to make sure everything went perfect for the party. When we had drinks the weekend before . . . they leaned into one another, his arm around her, and they were happy. Happy and relaxed. I remember thinking, *he's the one,* and I had felt this immense joy for my friend who, in many ways, was like a little sister."

IN BOZEMAN, KENNEDY AND PERKINS eagerly settled into their life together. That fall, they watched winter come early to the Gallatin Range. In late September the mountains outside Bozeman received a series of storms that dropped nearly six feet of snow, an occurrence that caught the attention of Doug Chabot and his small team at the Gallatin National Forest Avalanche Center. Chabot typically thought of this period as "preseason" for usual winter activities, like skiing, snowshoeing, and snowmobiling, activities that increasingly send people into the mountains as late fall and early winter come on.

In fact, Chabot sent his first advisory for the 2017–18 season on October 2, just after September's string of storms. "After the snowstorms that ended ten days ago," he wrote on the center's website, "I was wondering if we'd revert to an Indian summer or continue with a wintry pattern. Snowfall the last two nights points to winter, at least for now." In the prior forty-eight hours, Chabot reported that six to seven inches of fresh snow had fallen on Big Sky ski area, about an hour drive from Bozeman. He predicted another four to six inches for that night, and noted winds would increase to fifteen to twenty-five miles an hour. What caught Chabot's eye in particular as he reviewed weather data was that with the latest snowfall over the weekend of September 30 and October 1, especially strong winds had blown in the mountains. Now he saw them increase.

"Wind creates drifts," Chabot added in his advisory, "and when we have skiers hitting these drifts, we typically have avalanches. It is common to have avalanches with our early storms and I expect to hear of some in the coming days." So early was it in the winter season for Chabot, who often traveled out of the country in the summer, he confessed to "feeling rusty about avalanches, and imagine many of you are too. Over the years I have seen people make similar mistakes skiing early season so I lean on their errors to help scrub some of the rust away."

Chabot posted a litany of advice for anyone who might venture into the mountains in the early season. He reminded skiers to always have a beacon, shovel, and probe pole. He recommended helmets because of rocky areas that might still be exposed. He cautioned about "wind-loaded slopes" that may look "most inviting because they have full coverage." Chabot highlighted a single sentence: "Travel one at a time in avalanche terrain."

Early season snowfall creates an almost unbridled anticipation in a mountain town. While the advent of a new ski season may excite skiers who live in distant cities, for those who live in the mountains, the weather (and snow specifically) is a powerful presence, transforming the mountainous landscape beyond.

Posting early on the morning of October 7, Chabot characterized conditions in a "preseason" bulletin. The report stated: "Last night, the Bridger Range picked up six inches of new snow. This morning mountain temperatures are in the teens and ridgetop winds are out of the west at fifteen to twenty miles per hour. Snowfall is tapering off and the next shot of moisture is Wednesday night, which will be followed by sunny skies and seasonal temperatures." He wrote: "Ridgetop winds are westerly at fifteen to twenty miles per hour . . . and are strong enough to drift snow and create wind slabs. Areas with the deepest snow, least amount of rocks, and most inviting skiing will be wind-loaded areas: gullies and higher-elevation slopes. This presents a quandary because wind-loaded slopes are where someone could trigger an avalanche."

It's likely that Kennedy and Perkins decided before Chabot's October 7 advisory, or even earlier in the week, to hike into the mountains and ski that day. Whether they had followed Chabot's early season postings isn't known. Chabot and Kennedy were friends, partly as a result of Chabot's relationship with Kennedy's father, Michael, a world-class climber in his own right. Chabot had garnered attention in the alpine climbing world, receiving a 2013 "Mugs Stump" award along with two other climbers, a grant based on previous notable climbs and a source of funding for a future ascent in Pakistan.

WHAT IS KNOWN IS THIS: On Saturday morning, October 7, dawn broke in Bozeman with temperatures a little above freezing. The couple loaded their vehicle with backcountry ski gear: skis, climbing skins, backpacks. They headed for a spot called the Upper Taylor Fork trailhead, about fifty miles outside town. The trail is a six-mile hike approach to Imp Peak, an 11,200-foot pinnacle in the Madison Range. About twenty miles from Big Sky ski area, Imp Peak is the fourth highest mountain in the Madisons, presiding over an area known for grizzly bears, streams, and small alpine lakes.

Once Kennedy and Perkins reached the trailhead parking area at about 7,250 feet, they collected their gear and hiked up toward Imp Peak's north couloir. For the six-mile trek they climbed through fresh snow until the terrain began to steepen. Before them stood Imp Peak, rising like a snow-encrusted temple. With its three waterfall-like features and sheer slant, like that of a massive ski jump missing its launch ramp at the runout, Imp Peak's north face is imposing. On its perimeters the peak has two prominent shoulders that fall off sharply in breakneck routes. But the eye quickly confronts the middle slope. There the north face appears like the giant foreboding wall of a snow fortress.

The pair had reached just under ten thousand feet, facing the increasingly precipitous slope. As they continued up, there were visual and other signs to warrant some caution about what lay ahead. Kennedy and Perkins had each taken Level 1 and 2 avalanche courses, training that would have impressed upon them awareness of certain weather equations and terrain features to closely inspect or avoid. Up the couloir, the angle sharpened dramatically—nearly forty-five degrees—to a double black diamond ski pitch.

Even more dismaying was the wind and temperature that day. On Friday, a nearby weather sensor had recorded south winds blowing at twenty-five miles an hour, gusting to thirty-eight. The wind that morning was clocking steady speeds of thirty-five miles an hour and gusts on some surrounding peaks were reaching sixty-two miles an hour. The week's cold air at night had loosened its grip on the mountain, and sunny days had brought higher temperatures even into the lower 40s Fahrenheit. On this day they would reach 44 degrees. Such melt and freeze cycles are known to undermine snow stability.

Later, when Chabot dug a pit to investigate the layered snowpack, he reported finding that two previous days of south/southwest winds had loaded new snow onto an old layer from late September. As winds whipped the slab, it built in thickness and weight on the old unstable snow. Below, he reported, it had formed a "small grained mix of decomposing new snow and faceted crystals." Sitting on this like a vast white toboggan was the hard new slab.

By digging a snow pit like Chabot, a technique they would have learned in their avalanche training, the couple might have detected signs of decaying snowpack. But there was no indication one was dug. Had they followed a line well to the right or left of the main couloir, they might have encountered snow at a lower angle, ridgelines scrubbed to a harder surface by the wind. But they forged upward on what likely seemed familiar and untroubled terrain.

Long before this day, Perkins and Kennedy had learned a core lesson about the backcountry: mountains can magnify mistakes. Just a week earlier, Kennedy had published an essay delving into what he saw as essential questions about climbing. "Climbing is either a beautiful gift or a curse," he'd written. "I see both light and dark in climbing."

Within the bright capture of the couloir now, Perkins and Kennedy pressed forward. Tucked in the bottom of Perkins's backpack was her avalanche beacon. It was switched to "off."

HAYDEN KENNEDY HAD THE HANDS and feet of a prodigiously gifted athlete, steered by a mind that was incessantly searching. He was born into a family that was among American rock-climbing royalty. His father, Michael, served for twenty-four years as editor of *Climbing* magazine, then became editor of *Alpinist*. When not overseeing stories about climbing, the elder Kennedy made world-class climbs on such routes as Mount Foraker's *Infinite Spur* in Alaska. He climbed in Pakistan and Nepal, and took on American peaks in Colorado, Wyoming, and elsewhere. In the tight circle of American alpinists, Michael Kennedy was a highly respected mountaineer and a gatekeeper of climbing's legendary accomplishments. Julie Kennedy, herself an accomplished athlete, became the founder in 2014 of the 5Point Film Festival, a mountain film venue that shows an annual offering of works focusing on outdoor adventures embodying five values: purpose, respect, commitment, humility and balance.

At a young age, Hayden was exposed to experiences beyond just those that were athletic. He grew up climbing with his dad, as he recalled to friend Chris Van Leuven, who wrote about him in *Climbing*. But "I wasn't that into it," Hayden had told Van Leuven. Then, in his mid-teens, he developed a new interest in the sport. "When I was in high school . . . I started climbing with other people and not just my dad, and that's when I became my own climber." On a climbing trip to southern Utah with Van Leuven, Hayden referred to some of his earlier experiences on mountain faces, saying: "That was back before I was a climber."

Van Leuven and others saw Kennedy climbing in popular Colorado spots like Rifle and Indian Creek. The recollections were that Hayden "stormed up his pitches." But if he was displaying a dawning genius for the feel of rock, he was also forming a key personality trait that would only deepen in the coming years. As Van Leuven described it: "What I remember most about Hayden: his attitude. He had a hang-loose sincerity that stayed with you."

Hayden was a voracious reader when younger. "I spent my entire youth reading everything about our sport's history that I could get my hands on," he recalled in a piece for *Evening Sends*, an online climbing site, "not to mention discussing the nuances of climbing's ethics with my dad, a former world-class alpinist, and all of his fellow world-class alpinist buddies when they'd come through town." He had grown up around both climbing's overt physicality and its sobering literary epilogues. Luminary climbers were regular visitors to the Kennedy household; they were raconteurs and scrapers-by in some cases, men with thin bank accounts and monumental tales. There were personalities for the ages. Their stories and lessons infused the family's thoughts about climbing, especially Hayden's as he read their many accounts and aphorisms, such as Andre Roch's "the mountain doesn't know you're an avalanche expert"; Ed Viestur's "the mountain decides what you get to do"; and Conrad Anker's "the summit is what drives us, but the climb itself is what matters."

It wasn't just that Hayden wanted to escape from being assimilated into the soft-muscled backslaps of wider society; he would deem climbing a "rebel" sport, and his freewheeling lifestyle connected with that. It was as if he had decided to "send" (as climbers call completing a route) marquee climbs, but he resisted being automatically assimilated into climbing's pantheon. He didn't want to be blinded by media hype or other thin praise and lose the distance to examine with skepticism his motives. His evolving but unstated mantra seemed to be: make the climbs magisterial, but embody an of-the-people humility.

In 2013, *Outside* magazine had contacted Kennedy for a story on some of his increasingly notable achievements, but he rejected the offer outright. He told them, *Outside* later reported, that he was "not really interested in being in the mag. I have nothing against *Outside* but I would rather just be out of the media in general. I think that it distracts and over hypes everything, for me it's just not worth it. My passion for climbing is my own experience and doesn't need to be blown out of portion."

This theme was a recurring part of Kennedy's internal debate. "I don't hashtag on Instagram," he wrote in his *Evening Sends* piece. "I'm not on

Facebook. I guess that makes me a shitty millennial." It was a sentiment he continued to echo in other writings. "The bravado and the media hype of today are completely bullshit," he voiced in a 2016 essay in *Alpinist*. But his wariness about media coverage of his climbing went even deeper, touching his personal thought process about his own climbing and claiming summits that had gained the attention of others. "I also saw a side of myself I didn't like: bit by bit, I began to worry too much about what other people thought of my expeditions," he wrote in *Alpinist*. "I created an image of success in my mind that I couldn't live up to, a trap that didn't allow for failure or growth."

Hayden's father, Michael, became sensitive to this tension, and in a widely circulated *Alpinist* essay, "Letter to My Son," he addressed it head on. "When you head out in the future," he wrote, "other people will have expectations of you. Those notions will reflect their needs, desires, aspirations and fears. As best you can, clear your mind of the chatter. Don't think about how your life or climbs will look to anyone else. Make choices based on your values, your analysis, your intuition and your dreams."

As Hayden entered his late teens and early twenties, he had decided to put off attending college. Instead he took advantage of expedition invitations, and stitched together enough money to stay on the road from Colorado to Wyoming, Montana, and other destinations south. Some were spur-of-the-moment climbs, others elaborate trips requiring months of preparation. There was even a period when he was sidelined by a knee surgery and seemed to retreat into his earlier childhood misgivings about climbing. Of that period, he wrote in *Alpinist* that he "rarely held an ice tool or even thought about my alpine dreams." He picked up his saxophone again, which he had played in childhood, and began practicing. "I was following what I liked to call my 'no plan' plan: a mixture of general disorganization, a network of like-minded hooligans, a thirst for cheap beer and a sense of having nothing better to do with life."

But even cheap beer costs money. Taking on a grab bag of handyman-style jobs in Colorado, Kennedy and a partner opportunistically called one of those jobs their annual effort "to cash in on the joy." It was a full-fledged operation stringing Christmas lights on Front Range houses and the trees surrounding them. His ease with heights made the work appealing and allowed him to be inventive. "One time, when I couldn't find a solid roofing anchor," he recalled in his *Evening Sends* essay, "I filled my Grade VI haul-bag with bottles of water to create a counterweight on the opposite side of the house while I traversed a 40-degree slope of glaciated shingles. . . . I was so ready for a climbing trip when all this 'joyspreading' was done."

Kennedy reached a point, after his knee had healed and he'd pocketed money from work, when he asked himself, "Who am I without climbing? . . . Gradually, I forgot the summits that I'd reached and I remembered the friendships that I'd made." An answer to his question presented itself in 2012. Ambiguity about his life's direction and his disdain for media coverage had become powerful drivers, but even he could not have anticipated how the next expedition would change his life.

THE CLIMB, UNDERTAKEN WITH A PARTNER, Jason Kruk, was to be on the Southeast Ridge of Cerro Torre, an imposing 10,262-foot mountain on the border between Chile and Argentina. It's a region made famous in part by American climber Yvon Chouinard, who adopted an image of another local peak, Mount Fitz Roy, for the iconic logo of his company Patagonia. The face Kennedy and Kruk embarked on had first been climbed in 1970 by an internationally known Italian mountaineer, Cesare Maestri. Though he had made the summit, Maestri created what became a decades-long furor in the global climbing community. He had hauled up with him a three-hundred-pound, gas-powered compressor and a mining drill. As Maestri climbed, he drilled hundreds of holes into the mountain's face and fixed bolts in them that he used as anchors as he rose toward the summit. The passage was officially dubbed the "Compressor Route."

Kennedy and Kruk planned to climb the mountain by "fair means," not using any of Maestri's fixed bolts. Starting at a location called the Col of Patience, they climbed steadily. As they took on vertical pitch after pitch, they were revolted by all the bolts defacing the rock surface. They successfully reached the summit in a swift thirteen hours. Kennedy wrote: "After climbing Cerro Torre without the bolts, and seeing firsthand the outrageous nature of their placements—and understanding the history of the debate—Jason and I decided to remove a majority of the bolts on our way down."

They extracted more than a hundred bolts from the rock, a feat that made instant news around Cerro Torres's base camp. As word spread, they were viewed by some as new defilers of Cerro Torre's face, their actions seen as an attempt to erase a chapter in climbing's history. But many others, particularly the younger generation of climbers, saw them as heroes. They were praised for making a statement about returning the mountain to its natural state; they'd issued a call for others to scale the mountain as it is, not to rely on a lazy, forty-two-year-old route made failsafe with a tool that pulverized rock.

The debate broke out in mountaineering journals, on chat forums, and in other media, catching Kennedy off guard. A purer way of climbing and respect for mountains seemed a key element of the personal philosophy he'd been knotting together. It even had a basis in climbing's history: In Chouinard's early days at Yosemite, climbing El Capitan and other peaks, he had seen the rock scarring effects of pitons hammered into granite cracks. It was such an eyesore and gesture of disrespect for nature, Chouinard went on to design pitons climbers could remove. Later, he developed metal hexagonal "chocks" that could be wedged into cracks and easily withdrawn, leaving no trace at all.

Kennedy never credited some of his and Kruk's inspirations for their deeds on Cerro Torre to Chouinard, but as a student of climbing history, he was almost certainly influenced by what had happened in Yosemite. He found himself on the defense. The media attention he'd consciously dodged came steamrolling after him. Still in his early twenties, Kennedy found his father coming to his side, stepping into the glare of the debate with his powerful "Letter To My Son." "People will try to pigeonhole you with their words, but you aren't defined by what others think, only by what you know and by who you are, in your heart and mind. On Cerro Torre, you thought and acted with conviction and passion, making one of those decisive, spontaneous and honest gestures that can come only out of the uncensored soul."

Rather than silence or paralyze Hayden Kennedy's newly revived passion for mountaineering, the media scrutiny seemed to electrify his pursuit of it. In 2015, when he was twenty-five, that ardor was expressed in the name of a new mountain route, *Light Before Wisdom.*

THE COORDINATES OF THE MOUNTAIN—in the Kashmir region of India, in the Kishtwar Himalaya—were enough to incite a form of rapture. The 20,252-foot peak was known, in India and the surrounding region, as Cerro Kishtwar. Kennedy chronicled the expedition in vivid detail in a piece published in *Alpinist* in its summer 2016 issue.

On a trip to climb in Pakistan, Kennedy had met a Slovenian mountaineer, Urban Novak. They had hit it off, and in April 2015, Novak sent Kennedy a note asking if he would like to go on an expedition to India. He would join a team composed of Novak and two other alpinists, Slovenian Marko Prezelj and a Frenchman, Manu Pellissier. The men had picked Cerro Kishtwar's unclimbed East Face as their destination. Because of border conflicts with Pakistan, the Indian government offered limited access to the Kashmir region, but the team was granted approval and in the fall of 2015 they flew to New Delhi.

The trip got off to a less than harmonious flow. After eighteen hours of driving to a remote village in the mountains, the team took four days to transport gear, via horses and villagers, to their base camp. As the men began hiking toward their advanced camp, Prezelj began yelling at Kennedy: "Stop! Stop!"

Sporting headphones, Kennedy couldn't hear him, but then saw Prezelj dashing toward him and pulled down his headphones. "You motherfucker with your headphones," Prezelj screamed. "You are not in nature with that shit in your ears."

Despite encountering glacial run-off streams, meals of mystery meat, and cold energy bars for breakfast, the men soon melded like they had been climbing together for years. They exchanged leads on the icy pitches of Cerro Kishtwar, stopping at night on narrow ledges to try to get a little sleep. One night, using headlamps on their helmets, they just kept climbing. "I began to feel as if I were dreaming," Kennedy recalled in his *Alpinist* piece. "As the night sky unfolded, we became lost in the climbing. Each movement was slow. Cold and draining. Marko led pitch after pitch of calf-burning ice slopes. Our headlamps cast only small, spot-lit circles in the dark."

The next morning, Kennedy awoke to hear Pellissier playing the Rolling Stones's "Gimme Shelter" loudly on his iPod. After coffee and more cold energy bars, they packed up and headed for what they hoped would be their final series of pitches. They reached the summit around midnight, the sky clear and windless. "After all the days of climbing," Kennedy recalled, "we'd given everything we had."

Rising to the brilliant sun hours later, the men could see only vast reaches of "high peaks and deep valleys." Kennedy wrote: "Then the daylight strengthened like a rising scale, . . . notes of dark rock and white snow deepened. And for a second, I found exactly what I need out of climbing, something honest within myself." The name of the first ascent, christened *Light Before Wisdom,* spoke to the dawn light of the Kashmir range. But it also spoke to Kennedy's own inner ruminations of many years. He had seen the light of what felt like truth to him, but he was reaching for wisdom, for ultimate answers about his desire to climb.

Acclaim continued to follow Hayden Kennedy. He and the other three members of the Kashmir team were awarded one of climbing's highest recognitions, a Piolet d'Or. He had already received a Piolet for his 2012 ascent of Ogre 1 in Pakistan with partner Kyle Dempster. But following the high of *Light Before Wisdom,* life delivered news that again cast Kennedy into doubts about climbing. Two fellow climbers—Dempster and Justin Griffin, with whom Kennedy had climbed *Logical Progression* in Mexico—were killed in climbing accidents.

"It's not just the memorable summits and crux moves that are fleeting," Kennedy wrote for *Evening Sends* as he wrestled with memories of his friends. "Friends and climbing partners are fleeting, too. This is the painful reality of our sport, and I'm unsure what to make of it. Climbing is either a beautiful gift or a curse."

One thing that had always inspired Kennedy about climbing were its stories. In those, he now looked for new answers. "And maybe one genuine reason to try to share our stories about days we actually send something, when we are alive and at the height of our powers, is to try to bring back what's past, lost, or gone. Perhaps by doing so, we might find some light illuminating a new way forward." Beginning to glimpse that path, Kennedy explored an answer to his true motivations for climbing. It was in the people. "In many ways, I am still processing what has happened to my dear friends," he wrote in a piece for *Evening Sends.* "Waves of sadness overwhelm me at times, making it hard to stand up or focus. At other times, I am able to think only of the enchanting adventures, contemplative conversations, and the simple yet enriching moments we shared as friends. These pendulum shifts between various emotions will never go away, as I am starting to learn."

In 2015, Kennedy was invited to attend the International Climbers Festival in Lander, Wyoming, to give a presentation. Luke Lubchenco, who had attended high school with him, was in the audience. "He gave a speech about mountaineering to a room full of sport climbers," recalled Lubchenco in an article about Kennedy, "and he ended with three rules: 'Come home; come home with friends; come home with a summit. Do it in that order.'" Lubchenco called it "one of my proudest moments to be his friend." At the festival, Kennedy didn't just deliver a talk that left people thinking, as he had for years, about the purpose of climbing. He had met someone who would help him focus more precisely than ever on how important relationships were in his life: her name was Inge Perkins.

IN PICTURES, INGE PERKINS LOOKS petite, like a ballet dancer, a discipline she'd studied as a girl, or maybe a long-distance runner or competitive swimmer. What she pursued—like it was oxygen keeping her alive—was rock climbing and backcountry skiing. She thought nothing of acting on an impulse to hike twenty miles in the backcountry. She mountain-goated up cliffs that others placed belay anchors into. She skied steeps where mere experts would have preferred to descend with crampons.

Growing up in Bozeman, Perkins had come by all her mountaineering prowess honestly. She had been born to a Bozeman native, Steve Perkins, and a German mother, Heidi Hersant. The slopes and peaks of Bridger Bowl ski area had captured her imagination, and she used her strength and agility to become an adept black diamond skier. In her teens, Perkins worked at small Bozeman cafés and became well known among the town's residents. She had an engaging way with people a generation or two older than her.

When she graduated from Bozeman High School, Perkins was offered an academic scholarship at Fort Lewis College in Durango, Colorado. There she followed in her mother and father's footsteps by studying math. While her imagination had been fired by the ridgelines and peaks of the Gallatin Range, she had a penchant for mathematic's orderly formulas as well as the stray renegades like negative numbers. She found a perfect summation of this feeling in the novel *Smilla's Sense of Snow* by Peter Hoeg: "Do you know what the mathematical expression is for longing? . . . The negative numbers. The formalization of the feeling that you are missing something."

If Perkins was searching for what was missing in her life, she didn't wait for signs. When she saw things that spoke to her, she pursued them. On a family visit to Trondheim, Germany, when she was in her teens, she saw a film called *Wolf Summer* about a young female climber who scaled rock faces to save wolves. On her return to Bozeman, climbing became more than just her passion; it seized her mind as much as it did her hands and feet. Perkins's sparrow-like physique lent itself so naturally to climbing it seemed as if, rather than climb a rock face, she could just wing her way up with a few quick flits of her arms.

While still in college, she received an invitation to become an au pair for a German family in the Bavarian town of Bamberg. In the mountains there, she continued to run, work on climbing and skiing, and perfected her German. Returning to Bozeman a year later, she enrolled at Montana State University and resumed her math studies. This allowed her to reconnect with her family, especially her father who had a zeal for being in the mountains, particularly backcountry skiing. Family members recall how Perkins and her father would go into the backcountry and "tour together, sharing their love for the mountains with one another as they skinned up peaks and carved through powder." At Bridger Bowl ski area, she took her mother "through ridge shoots and forays into the side country."

In her late teens, Perkins began to be noticed on the competitive climbing circuit. She worked out at climbing gyms, Power Company Climbing in Lander and The Spire in Bozeman. Perkins advanced to harder climbs and picked up sponsors such as Scarpa climbing shoes, Mystery Ranch backpacks, and Petzl climbing gear.

Among her friends, Perkins was like morning alpenglow, warming and inspiring them with her presence, not because her personality was always electrifying—though it could be that, too—but because she turned her warmth on *them*. Countless friends point to her focusing a light on them as she elicited details about their lives, feelings, dreams. "Inge had this deep way of listening that spoke of an older-than-her-years way of moving through the world," says Kelsey Sather. "She would sit in a conversation and look at you as you talked, her eyes becoming distant only to consider what you had said. This act of being present with another person, truly *present,* is not something easily achieved, not frequently encountered."

But as much as friends mattered to Perkins, she had a vagabond streak. It prompted her to jump into the driver's seat of her truck, which she had named "stallion," and head off to Lander, Wyoming, or Jackson Hole, or even points east, like West Virginia. Armed with postcards, lavender-scented cotton balls for her truck's dashboard, and all her meticulously selected and cared-for climbing gear—harness, slings, carabiners, shoes with pliable soles and uppers—she would pick a destination and just go.

Yet while her peripatetic time on the road could be lonely, she would invariably land in the company of a friend or family member, someone to whom she could, like a climber scaling higher, clip her life's harness. Her travel and training had begun to pay off. She joined the Bozeman Climbing Team to compete in regular competitions. She won the Montana Bouldering Championships, a low-altitude form of climbing, and the Montana Randonee Championships for free-heel skiers. During the 2015–16 school year at Montana State, she managed to create so much velocity in her climbing ambitions that she climbed in seven states and five countries. The names of the climbs spoke of her world: *Rodeo Free Europe, Manhattan Project, Vesper, No Country for Old Men, Roadside Prophet.*

Her name was becoming recognized in the American climbing community, but like Hayden Kennedy, Inge Perkins found her inspiration in the climbs and in friendships, not in accolades. In a February 2017 online interview titled "The Work Behind the Body: Inge Perkins," Sather asked Perkins about her motivations in climbing and other sports. "It can be easy to begin to be influenced to do things that will seem 'badass' and marketable but do not align with your personal objectives and direction," Perkins told her. "That is when the sport, or more like lifestyle, can lose such precious meaning to the individual. And that personal relationship with outdoor pursuits such as skiing and climbing is so important for staying present in your surroundings and making the best choices possible."

Like her impulses for road-tripping, Perkins didn't dismiss the occasional impromptu need for just letting her mind and body go. "Don't be afraid to let it all loose on the dance floor. It is the most wonderful way to forget any notions of others judging you."

AS INGE AND HAYDEN SKINNED up toward the crest of Imp Peak that Saturday morning in 2017, the crest stood a thousand heavenly feet above them when they saw the avalanche break. The crown, which spanned 150 feet, left a jagged scar across the high slope as it ripped free. Three hundred feet from top to bottom, the avalanche was wind slab snow created by a night of powerful gusts heaping dense flakes into batches of snow cinder blocks.

As the avalanche gained speed, the snowpack ruptured all around them. The heaving blocks looked like pieces of a dynamited building coursing down a slope. In the moment it all happened, so impossible was the world before them, a world that a second ago had been sunlit and beckoning, it must have seemed an hallucination. But there was no time to fall back on formal avalanche training, or even flee. When the impact hit, it bore the punch of a tidal wave. Skis that moments earlier had been so rhythmic under their feet likely became anchors dragging them under. Knocked backward, they might have instinctively tried to swim toward the sky. But the snow swallowed them, a cold riptide running downward.

Chaotic and disorienting as the first moments were for Perkins and Kennedy, the slide did not last long. When it stopped, the entire debris field, a vicious battleground of collapsed snow, had only traveled 175 feet. They were caught within 30 feet of each other, barely 100 feet down from where the top of the avalanche had snapped free. Yet 30 feet apart may as well have been the full reach of Imp Peak's 11,200 feet. In a fraction of a minute, everything on the mountain fell into an astonished silence. Motion ceased. A thin vibration of violence hung on the wind. Only the wind, quivering across the snowfield, uttered any sound.

The mountain tells you what you get to do....

WHEN HIS INITIAL PANIC HAD passed, Kennedy realized his head and midchest were above the snow. Frantically, he began to dig. Minutes passed as he cleared snow and forced himself free. Released from the heavy debris, he almost certainly cast a visual search over the slope, looking for a sign of Perkins, for some swatch of color in the white sea before him. Finding his avalanche beacon, he switched it to "search" and waved it over the snowfield's jumble of broken snow. His beacon scoured the mountain's flanks, finding nothing it could connect with, no radioed impulse.

There could only be one thought now: *Inge*. Heart pounding, he grabbed his shovel and probe pole. The terrain around him, naked in full sunlight, itself seemed shocked. He looked for a clue that might guide him to begin probing and digging. He tried to pinpoint where they'd been standing when the avalanche hit. He sought to gauge how far she might have been swept away.

Descending a short distance, Kennedy stepped in the rubble of snow, slow, unstable. It snagged on his boots, but he ignored the slippage and lurched forward. Down and down and down, he drove the probe, forcing it through a bottomless barrel, or feeling it strike ground beneath the snow. He'd practiced this in training. He knew how to do this. Down and down and down and down.

Winds raked the eleven-thousand-foot peak above him. As a graduate of Avalanche 1 and 2 level certifications, he knew well the tenacious, stopwatch math behind burials. If he was going to uncover her in ten or fifteen, or even thirty minutes—the last at the outer limits of survivability—he had to expend every ounce of physical exertion he had. Digging and covering distance in thigh- to waist-deep snow is, even for the fittest person, an activity that becomes anaerobic. You go past burning calories to burn muscle from the fiber of your arms and legs. Your lungs begin to ache as if a caustic chemical has been sprayed inside them.

Thirty minutes passed, but Hayden kept going. *Inge.* An hour and he was still frantically hunting for her. *Inge.* Time had all but run out. Barely an hour ago, he could not have imagined this. In this moment, Kennedy could not have been more alone if he had been drifting into space. In the solar oven that the mountain face had become, he persisted in probing and digging. Strong and agile, he drew on every fragment of strength that had taken him up thousands of feet of granite. Sweat poured from his skin. But he didn't find her.

A second hour passed. He probed and dug. Near the end of the third hour, Kennedy had burned through all hope. He knew she was gone. *Inge.*

Out there, stranded on the slope with only despair to guide him, Kennedy made a final plant of his shovel and probe into the heart of the avalanche. It would help them find her, bring her out. This wasn't the place he could allow her body to lie. He took one last look.

There was no immediate exit from the mountain, no way to escape the unendurable crush of emotions he must have felt. Psychologically, the three hours of focus and calculation he had relied on to try to save her melted into a dark, haunted corner of the mind. He turned and began a six-mile hike out. Alone.

IN THE HOURS TO COME there were echoes of her life everywhere. In the drive back to Bozeman, in the rooms of their apartment, in her gear, in her clothes, in her math books. All those whimsical and precise pieces of the life of Inge Perkins, silently raging against her death. Somewhere among the walls, the scent of lavender lingered.

In the midst of these echoes, Hayden Kennedy composed a letter. He described exact GPS coordinates for the location of the avalanche on Imp Peak where Inge had been buried. He said he had left his shovel and probe pole in the snow where he'd searched. He wrote words he intended for no one other than his mother and father. His life was his life, and the road ahead must have looked infinite, the time a never-ending reckoning. Every minute, every hour, there would be the possible specter of that October morning on Imp Peak. A day he might never allow himself an excuse for.

Hayden Kennedy had been striving to go beyond enlightenment, to touch wisdom, if only to gain glimpses of the answers he sought through his writing and dialogue with the wider climbing community. The world has seen famous athletes who never venture deeply into the questions he struggled with, much less do it in public. He had come far closer to that wisdom than many his age had a right to. His reflections remain. "For a young man who is at the top of his game, arguably one of the top alpinists in the world," says Doug Chabot of the Bozeman avalanche center, "to have those kinds of reflections and put them out into the public eye is unique."

Hayden Kennedy finished his letter, so clear in its instructions guiding others that it was as precise as if he had been leading a dangerous climb. Light before wisdom. Precise. As a climber, he had always embraced that quality. He must have wanted his last minutes to show how much it meant to him. She had been a mathematician. The precision of numbers mattered to her.

Finished with the letter, he placed it where it could easily be found. Sometime between that night and Sunday afternoon, he released himself from the world. He did it as surely as if he'd been unbuckling himself from a climbing harness high on a rock face.

THAT SATURDAY NIGHT, PEOPLE IMMEDIATELY recognized it was unlike Kennedy to not show up at a party he had been expected to attend. At first, they worried. Then they drew up likely explanations to dilute those concerns. The next afternoon, Kennedy failed to show up at Blackbird, the pizzeria where he worked.

After a day and a half of friends trying to contact the couple, Perkins's stepbrother, CJ, went to the apartment that Sunday night and made his way inside. There he found Hayden Kennedy's body. He summoned the Bozeman police, and shortly after arriving, they found the extensive note Kennedy had written.

At about 10:00 p.m. that night, a call went out to Gallatin County Search and Rescue, notifying them of the avalanche accident. Soon, Doug Chabot was brought into the loop with police and the search-and-rescue team. "The next morning, at first light," Chabot says, "I flew in. I was on the first ship to the site."

Along with Chabot on the helicopter that morning were two members of Gallatin County Search and Rescue as well as an avalanche search dog. The pilot followed the coordinates Kennedy had detailed in his note. It was about 10:00 a.m. when the helicopter's rotors beat the air above the approach to Imp Peak. The sun was out, snow shining below with a placid luster as if ready for an unblemished day. Belted into the heli, Chabot forced himself to detach emotionally from the image of Inge Perkins in his head. He had work to do and wanted to be as professional as possible. The job before him was to find her body and bring her home to Bozeman.

From the air, Chabot and the others could immediately see the avalanche debris as they came within range of the peak. Slightly off to the left of the couloir's center, the river of toppled snow blocks looked like large, frozen whitecaps. They could also see the brightly colored shafts of the shovel and probe pole marking the spot where Kennedy had left them.

At 10:20 a.m. the helicopter ferrying Chabot, the dog, and other searchers settled down beside the debris field. Chabot inspected the spot where Kennedy had been partially buried and had dug himself out. Close by were signs of his search for his partner.

"You know it makes sense to me," Chabot recalls of the probe marks. "If you and I were skiing together and we both got caught in an avalanche, and I ended up in my spot and I dug myself out and I didn't know where you were, chances would be pretty good you would be near me . . . That made sense, looking at it." After letting the search dog cover the immediate zone around the hole left by Kennedy, Chabot expanded the radius of his own probing. While he worked, an incident command center was set up at a nearby trailhead, Wapati Creek. At 11:00 a.m. a second helicopter dropped three more searchers and a second rescue dog. Within ten minutes, a third team of three more searchers was deposited on the mountainside.

Even though Kennedy's note said that he and Inge had not done a beacon check before heading up the couloir, the teams performed avalanche beacon canvasses, continued to probe, and worked the two dogs over the debris field. Up the hill from where Kennedy had pulled himself out of the snow, Chabot probed. At 11:22 a.m., just twelve minutes after the third search team had arrived, he felt a firm halt of his probe pole about three feet under the snow. The hole around which Kennedy had sought Perkins's body was thirty feet down the face from where Chabot now stood.

Carefully they dug out her body. It was not the first time Chabot had helped uncover a deceased victim buried in an avalanche, nor would it be the last death he'd deal with in that ski season. But this was different. He knew Inge Perkins, not well, but he knew her as a Bozeman native and "a badass" climber and skier. He knew she and Hayden Kennedy had been romantically involved. Still raw in his mind was the news of Kennedy's death.

When the snow had been cleared away from her, Chabot covered her face "to keep the recovery impersonal." He measured her depth in the snow, noted her body position, and took some photos for his investigation report. "Lastly," he says, "we took off her skis, straightened out her body." Chabot had the dreadful task of helping wrap her lifeless body. The procedure had to be done with a care that displayed respect for her. "During the entire recovery, we were quiet and somber," he recalls. "Many of us knew her."

In minutes she would be evacuated from Imp Peak on a helicopter longline dangling her body above the cold, bright air hugging the Madison Range peaks. At 11:44 a.m., less than an hour and a half after the initial landing, the helicopter delivered her to the Wapati Creek command location. From there, she was driven to a small hospital in Big Sky.

As avalanches go, this one was not large. But it was heavy and had been initiated on very steep terrain, which meant it had thrust itself forward swiftly and bore an inflexible punch when it hit. "It was about as small as you could have and kill someone," Chabot says. In his accident report, he characterized the Imp Peak slide as HS-ASu-R3-D2-O—a detailed system of acronyms avalanche forecasters use. Chabot's gradations meant the avalanche was "hard slab/artificially triggered by a skier (unintentionally)/rated 3 on a scale of 1–5 with regard to how much of the path the slide's width filled/destructive at a rating of 2, on a scale of 1–5, with 2 strong enough to kill, and 3 to 5 powerful enough to damage houses and break trees in half/O for old snow as the underlying cause."

After completing his investigation in the shadow of Imp Peak, Chabot went through Perkins's pockets to see if he could locate an avalanche beacon. No device appeared. Later, when a coroner was performing an autopsy on her, the examiner found the beacon. "Her beacon was in the bottom of her pack," Chabot says. "It was fully functioning, but it was off."

Thinking about the climb itself, Chabot says, while it was steep and certainly required experience in the backcountry, it wasn't extremely technical. "They weren't putting up a first ascent. It was a pretty benign approach to a peak." But then, Chabot, who has climbed big walls in Alaska, India, Nepal, and Pakistan, pauses, and says: "I try not to forget that for myself." He continues: "Looking at the avalanche, what strikes me, again and again, no matter how accomplished we are as climbers, and skiers, outdoorsmen, when we go out and we enter avalanche terrain, there's an element of risk that we need to be aware of. And that avalanche, and small terrain trap will kill you."

While Chabot had a professional duty to objectively handle the avalanche investigation, understandably he could not separate his feelings, beyond the report, from memories of his friendship with Kennedy and insights into his character. "For Hayden, the grief that came from feeling helpless, was very real. And Hayden was not a depressed person. He was a very upbeat person. He always saw the best in things, he always saw the best in people."

Chabot still struggles to make sense of it. "It makes me think, my God, if Hayden, if he can be dragged down so quickly and thoroughly in grief to where he'd commit suicide . . . then that could happen to anybody. I'm not immune to that, I'd be foolish to think I'm immune. That's a wake-up call. That's something I don't think about often. Knowing or seeing the aftermath of that, definitely it's like, 'Wow, we're fragile beings.' Our minds are fragile. We need to acknowledge that. . . . Someone so tough can still be taken down by these very real human emotions." Remembering that morning on Imp Peak, finding Perkins's body, Chabot says: "This avalanche scarred me like no other."

23

CHASING DOWN THE DRAGON

In early 2018, three months after the Imp Peak avalanche, forecaster Craig Gordon puzzled over the winter's lack of snowfall and what secrets might be hidden in the snowpack. In his mind, where the lifespan of a season's snowpack ranges from infancy to elderly, the age of the January snow was still in its rebellious, potentially wrathful teenage state.

Though it was mid-January, he observed: "The snowpack thinks it's November." To Gordon, this was not a good thing. Lengthy stretches without fresh snowfall meant the existing snow was going through the equivalent of a metastasizing cancer, its structure riddled with decay. "This season we have been super lucky," he said, "because we've dodged so many bullets, so many close calls. The only thing that's kept us from not experiencing a backcountry avalanche fatality is the low amount of snowfall. . . . Every time we load snowpack up with water and wind, the weak layers in the snowpack get reactivated."

While there had so far been no avalanche fatalities in Utah, Gordon knew of Inge Perkins's death and the subsequent death by suicide of Hayden Kennedy. The story had left a surreal feeling about the winter's beginning. And by late February there would be another four skiers and one snowboarder dead from avalanches in Alaska, Colorado, Montana, and Wyoming.

It wasn't as though avalanches had not been breaking in Utah. There had been scores of them. Just none that had taken under skiers or snowboarders and killed them. The strange conditions in the mountains had bred a kind of foreboding, rather than excitement, about the arrival of new storms. Loading weight and water onto the volatile, water-perforated snow would only bring problems. With the approach of each stormfront, danger seemed to whistle in the wind.

And then, on Sunday, January 21, two skiers were caught in an avalanche. Inside a ski resort. It was just after 3:00 p.m. at Snowbasin resort overlooking the Ogden Valley. The day before, a colleague of Gordon's, Drew Hardesty, had posted an advisory at 6:00 a.m. A storm had dumped eight to twelve inches of new snow on the old, icy snowpack in the mountains above the valley. Hardesty noted that no avalanches had been reported the day before, a Friday. But he added "during this last avalanche cycle, eight people were caught in avalanches with several close calls." Sunday broke with the sun fighting to shine through "the last few flakes of cold smoke from the departing storm," as Hardesty described it. He forecasted at 7:24 a.m. that the day would be "the stuff that dreams are made of."

Skiing in-bounds in an area of Snowbasin called No Name, a twenty-three-year-old woman triggered an avalanche and was carried down the mountain an estimated one thousand feet. Her thirty-five-year-old brother was also caught. When the slide stopped, the woman was completely buried. Her brother was only partially covered. Fortunately for them, another skier and a Snowbasin ski patrol member saw the avalanche and were able to quickly reach the woman, who was buried under four to five feet of snow. Within ten minutes she had been dug out. A LifeFlight helicopter flew her to nearby McKay-Dee Hospital in Ogden, and both the woman and her brother recovered.

The slide, which received wide media coverage in the Salt Lake Valley, particularly because it was in a ski resort, spooked people. If an avalanche like that had occurred in heavily controlled, in-bounds terrain, people wondered what might be concealed in side-country at resorts or, more alarmingly, in the backcountry. The day after the slide, on January 22, Gordon filed an advisory: "There's a few variables out there right now, but what I do know is . . . we have a complex snowpack and this season's history reveals that each time it snows, weak layers in our mid-pack wake up and we see a string of human-triggered slides."

Suspicions about the snow began to creep into other forecasters' avalanche advisories. The Friday after the Snowbasin accident, January 26, a colleague of Gordon's, Greg Gagne, posted a warning for the Ogden area mountains in which he said of the snowpack: "I just don't trust it." Gagne wrote: "Persistent weak layers can be found down one to three feet, and yesterday's strong west/southwest winds, as well as new snow (with nearly 0.5 inch water in places) have put a new load onto these slopes, stressing these weak layers once again." The next day, Drew Hardesty posted the Ogden area advisory, observing: "These avalanches are insidious: they continue to occur with many tracks already on the slopes, with no signs of cracking or collapsing, and each often triggered from a distance."

Standing watch over the Uintas, Gordon saw even more disturbing evidence of avalanche activity as January ended and February began. By February 19 he posted a forecast that said avalanche danger was "considerable." Two weeks later, on March 4, he posted another "considerable" warning with the added observation: "Heads Up. . . . If winds increase or we receive more snow than forecast, the avalanche danger will rapidly rise to high."

In some ways, Gordon's long tenure has led him to feel even more acutely the weight of trust people put in him. "At first, I thought it was just the ski and snow community. But you realize you can touch somebody's life, you can make an impact. . . . It can be the person who watches the quirky avalanche guy on TV, and then tells his neighbor, who's loading up his snowmobile on a Friday night, 'Hey, this guy says it's dangerous.' That's where the deeper connection is. It's such a blessed life. I truly believe that. . . . This arc of connection, you don't realize it until somebody tells you how you've impacted their life."

Jittery sensations that winter about sporadic snowfall weren't limited to Utah. Gordon's March 4 forecast had come a day after an enormous avalanche closed down the entire Mammoth Mountain ski resort in California. The day before that, another massive slide had hit the Sierra's Squaw Valley. The Mammoth slide had happened on a crowded Saturday morning, just after 10:00 a.m. Descending from near the eleven-thousand-foot summit of the mountain, the avalanche came rocketing down Climax, a steep, expert run. Producing a huge powder cloud that could be seen by hundreds of skiers and boarders, the slide ate up so much in-bounds terrain that ski patrollers and others feared multiple people had been buried.

"It was pandemonium everywhere you looked," Los Angeles resident Barbara Maynard told the *Los Angeles Times*: "Ambulances, police vehicles and fire engines were rolling into the area. Simultaneously, Mammoth Mountain staffers and ski patrols were roaring up the slopes on snowmobiles." As skiers fled the mountain in cars that backed up in a long line, they passed "more than fifteen ambulances, their sirens screaming," reported the newspaper. No fatalities or serious injuries were ultimately reported. The newspaper captured some of the same edginess in California skiers that those in Utah had experienced. "The varying consistency of the snowpack deposited over the area by recent storms," the news account said, "already was a topic of conversation among locals concerned about the potential for avalanches."

At Squaw Valley, the day before, an avalanche had swept over two men and three women near a chairlift. Nearly a hundred rescuers with probes and search dogs had been brought in to comb through the snow. No one was found buried. But to add to the high anxiety, the Squaw Valley avalanche came just a few hours after the body of a snowboarder, who'd disappeared on Thursday, had been dug out of the snow.

IT WAS EASY TO BE rattled by all this. I met up with Gordon one afternoon to gauge his feelings about the ominous undertone that discussions of ski conditions had undertaken. "In one regard," he said, "getting nickeled and dimed by storms doesn't provide enough strength to stabilize the snowpack. The other side is we haven't gotten walloped, in which case the avalanches would break bigger and deeper."

Gordon's biggest worry was that there was so much pent-up energy to ski the forthcoming classic Utah spring conditions. If a big storm rolled in, people might ignore the avalanche risk and just let loose. "The lack of volume has actually helped us get away with a lot," he said. "If we can just ride off into the sunset and wash our hands of this season, we'll be okay." But there was still at least another month left in the ski season. And Gordon, as he always does, was fixating on where the white dragons could be lying in wait. "It could get sketchy," he said.

Just two days after the Mammoth Mountain storm and avalanche, a legacy of the California storm passed through Utah leaving new snow across the ranges. On Monday, I skied with my wife and older son and his fiancée at Sundance ski resort. That morning, Greg Gagne had issued an advisory for the Provo area mountains where Sundance sits, noting avalanche dangers were "considerable in upper elevations." The advisory warned: "With what many are describing as the best conditions of the year, it is easy to let the brilliant riding cloud good judgment."

The next day, we skied Powder Mountain in the Ogden Valley. From the beginning of our runs, we heard avalanche bombing in the mountains to the west of us. The morning's advisory had noted: "Yesterday one ski patrol in the Ogden area triggered an avalanche almost big enough to bury a car." We were repeatedly riding a chairlift called Hidden Lake, and from the chair I heard the sound of more bombs going off. In that direction were two well-known, snowcat-accessed ski descents, Lightning Ridge and James Peak, the latter a 9,421-foot summit I had once heli-skied.

I focused on the distant, rugged triangle of the mountain's apex. Then another bomb exploded near the top. From where I sat on the lift, I saw an avalanche gathering speed as it grew into a colossal white fist, a cyclone battering everything in its path as it spread down the mountain. I had never before seen anything like it. By the time it reached the runout at the bottom of James Peak, I had one prominent impression: nature's sheer, unconstrained *power*.

The next morning an advisory read: "Explosive work with avalanche reduction resulted in a couple very large avalanches in the Powder Mountain area yesterday. These were two to five feet deep—some stepping to the ground—on steep NE and ESE facing slopes above nine thousand feet near James Peak."

I remembered Craig Gordon's words: "It could get sketchy."

24

ESCAPING THE DRAGON OF THE UINTAS: UTAH'S JEREMY JONES

More than a year after he'd been caught in an avalanche in the Uinta Range and had both his legs broken, Utah snowboarder Jeremy Jones was still limping, had ventured over the winter just a few short runs on a snowboard, and was living every day for the next one. "Tomorrow's always the day for me," he told me. "Tomorrow will be better."

It was mid-April and a late winter storm was pummeling the Wasatch mountains when I drove into Salt Lake City to reconnect with Jones. Last time I'd seen him he was balancing on crutches, his leg bones just beginning to mend from the mid-January accident. Now, as he walked toward me inside a brew-pub, the crutches were gone but he had a small hitch in his gait.

Outside the telltale consequence of his accident, Jones still had the persona of an off-duty pro snowboarder. He was dressed in baggy jeans, a trucker ball cap, and a hoodie bearing an image of a bucking bronco mounted by a rodeo rider and a slogan from a Wyoming business: "We're No Cowboys." Jones said he had seen his life go off a snowboarder's equivalent to launching a jump from a kicker landing hard. Other times, he hit smoothly and felt better days were ahead.

"Sometimes, it's uncomfortable," he said, thinking back to the avalanche. "You're not ready to visit it. You want to close the doors on it, and lock it in the safe. . . . But you just have to push past it." One of the hardest aspects had been his slow physical recovery. He had stainless-steel rods in both legs, and a steel plate attached to his right fibula. When he had an operation about six months earlier to remove eight screws from his right leg, the surgeons found that his bones weren't healing as expected. They performed a bone graft, which meant he had to limit his activities again until that healed. It was a setback Jones hadn't anticipated. "I came out of the operation thinking I was going to walk out," he recalled, "but I had a boot on."

Yet he's had glimpses of tomorrows that are both bright and meaningful. In late 2017 he was invited to give a presentation in Nelson, British Columbia, to seventy young snowboarders. They were there for four days of avalanche training and boarding in the backcountry. Jones gave them a first-person account of being caught in an avalanche.

"I know how many years we skirted this," Jones told the audience of his narrow brushes with avalanches, "but [you] need to know what it's like in the eye of the storm." Standing before the young gun snowboarders, Jones could see their reaction to his story ripple across the room. "You see the impact immediately, these young kids," he said. "Listening to me, it impacts them and I hope that turns into something life-saving."

One thing he relates to people is his belief that, despite the accident, his crew of fellow boarders acted effectively when faced with a rescue. Rather than just a tale of the avalanche itself, he said, "it's turned into more about . . . how my boys behaved." He explained: "Those are the guys that solved the problems once nature came. . . . They did what everyone wants to pull off." This British Columbia talk was part of a formal program, so Jones was prepared for it by the time he arrived. But not all his interactions are as smooth when he talks about the avalanche spontaneously, or even when he encounters some of the boarders who were with him on the snowcat that day. "Some people seem to feel you don't want to talk about it," he said. "For me it was good, it's what I needed to do."

With those who were stranded with him on the mountain that day, he said it's as if the memory has led to an unspoken pact. "The bond that incident created is strange," he said. "It's kind of an intense meet-up. You feel your emotions spike. It's kind of cool."

The healing regimen Jones has been on is one his wife and children have shared in. Many days, he retreated alone to his garage to lift weights and perform other rehabilitative exercises. "I couldn't focus on my family the way I wanted to," he said. At first, his kids had a hard time emotionally adapting to the struggles they witnessed their father going through; he had always presented such an image of strength before the accident. "They had to mentally rehab too," Jones said of his kids. "A year and a half is a long time for a kid." Over time, he realized that what became important was for him to show his family how determined he was to recover from the accident, physically and mentally.

"I feel like that lesson I'm trying to teach my kids is if I died in a year and I had hung it up [on trying to recover], I never would want to put that imprint in my kids, it's not the imprint I want to leave. That became really, really good fuel to keep going." Visualizing his image in his kids' minds, he said: "That's how I picked myself up off the floor everyday, literally. They need to see me standing."

When he was in British Columbia speaking to the young snowboarders, the snowcat guides offered Jones a chance to get back on a snowboard. Still cautious after the bone graft operation, he decided to try a few slow runs on low-angle terrain. There was powder snow to soften any fall he might take, and when he came to cat tracks or bumps, he unstrapped from his board and carried it. While his few runs were short, for Jones it was more about getting over two potential barriers: the physical one and the emotional one of being in the backcountry again. He felt like he was well along in conquering the physical one. And he was clear about his intentions for the backcountry.

"When I talk to people it's one of the first things framed up as a question," he said. "'You wouldn't go into the backcountry again?'" His answer: "Absolutely. I might not work as close to the edge. A lot of that has to do with age and avoiding injury. I have to recognize that." Of his snowboarding future, he said, "I'll never be what I was. That's not the point. The point is to not let this break me."

If the avalanche accident forced Jones to deal with surgical hurdles, the tedious repetitions of rehabilitation, family sensitivities, and public perceptions about his post-accident state, he also had to ponder large questions about why life threw him such a curve. Questions about danger and death in the mountains. He doesn't dwell on the why-me dimension of being caught in the slide. "I don't really visit that. To me it feels like a waste of energy. It would be speculation and would be that forever. . . . I didn't want it to cripple me, physically or mentally."

As for dangers in the backcountry—and the ever present specter of avalanches—Jones is at peace with that. "I long ago accepted if death in the mountains is how it goes for me, I'm okay with that." It's a philosophy he hopes will extend to his kids someday—a healthy respect for the mountains but not a fear of experiencing the beauties of the backcountry. As his kids get older, he hopes they'll project such a mindset: "I still want to go to the mountains, but I'll remember what my dad went through."

OUTSIDE THE BREWPUB, A LIGHT rain was falling in the Salt Lake Valley, but the Wasatch Range was hidden in ashen clouds as snow fell at higher altitudes. Forecasts were for as much as nine inches of new powder. Looking up toward the storm's drawn curtain, Jones might have recalled the day he and eight other snowboarders made their way, in a blizzard, into the Uintas.

Zipping up his "We're No Cowboys" hoodie against the chill, Jones turned to me and said he thought the weather might clear by morning. He was thinking about going up to his home mountain at Brighton ski area. If so, he might strap in for some end-of-season rides. But then, qualifying it, Jones said if he went, he would stick to groomed runs.

I could tell he was not just thinking of the fresh snow at Brighton, but also imagining the cool, whitewashed backcountry beyond it. It was one of his tomorrow moments. He smiled broadly as he considered the prospect of strapping in again. Progress toward next winter, and a return to the backcountry, would remain a solitary pursuit for now.

Tomorrow's runs could be another step toward leaving behind his memories of the avalanche.

25

SURVIVORS: PHANTOMS OF THE DRAGON

Six months after being buried in an avalanche in British Columbia, Jared Flitton had to have his ACL and MCL knee ligaments repaired from the torn muscle he suffered in the slide. Waking from anesthesia after the surgery, Flitton felt a surge of panic in his chest as he gulped air, disoriented, certain he was trapped inside a coffin of snow. "As I'm coming out of surgery, waking up," Flitton recalled, "I started freaking out . . . I thought I was buried again." The panic attacks didn't end there. Flitton would wake up in the middle of the night "feeling like I was buried in snow, throwing the covers off because I thought they were snow."

As a dentist, Flitton has to display a certain calmness and reassuring manner to his patients. But there were days, he said, he could barely contain his anxiety as memories of the avalanche intruded. It seemed the only way he could force away images of the burial was by calling one of the other men who also had been caught on the mountain that day. "If I was having a rough day," he recalled, "I'd call one of them and say, 'The avalanche is wigging me out.'. . . It was harder for me mentally than physically."

Some accounts of the aftermath for avalanche survivors compare it to post-traumatic stress disorder (PTSD), as if the survivors came back from a war zone. When someone emerges from the experience as a survivor, while someone else has died, they describe feelings of survivor's guilt—why did I get to live?—and a sense of being under scrutiny from others as if they made a mistake that killed another person. That can lead to a deep immersion in shame over an accident that may have been beyond the skier's or boarder's control.

Therapist Jennifer Fiebig has treated survivors of avalanche accidents. A Bozeman resident at the time of the deaths of Hayden Kennedy and Inge Perkins, Fiebig told me that survivors are often the ones who receive the least help while recovering. "There's a culture of analyzing all the data after an accident," she says, "that takes all the humanity out of the survivors, who are already thinking 'I killed my friend,' or going through other grief, and then descending into a shame spiral on top of that." Now working in Durango, Colorado, Fiebig said there was a psychological "shockwave" that went through Bozeman when people learned of the loss of Perkins and Kennedy. "I have friends who were very close with Inge. There's still shock and grieving in the community. . . . I wish there was more space in the community for people to talk about it."

Jake Rabe, who was also buried by the avalanche in British Columbia, was resuscitated just feet away from where Flitton was buried. He went through major stages of recovery in the year and a half after the accident. "There was a lot of anger initially," he said. "At least in speaking with other people who've had life-altering events, your life feels like it's in standstill and you see everybody else continue moving around you. You almost get angry, especially with social media. I would see people going on vacations and I'd be like 'What are they doing?' I'd have to remind myself people still have their own lives and they're still continuing because it just felt like we were in this huge standstill, physical therapy and everything else."

For the first year after the accident, Rabe said, the weight of his injuries bore down heavily on him. The severed nerves in his right arm and neck left everything below his shoulder paralyzed. It seemed he had barely finished eleven years of medical school and a demanding hospital residency, and now he was unable to work. "I have an enormous amount of student loans," he said of the added pressure he felt, pressures that weighed just as heavily on his family. "Our life was where we wanted it to be," he said. "And then all of a sudden that was all gone."

After being told by local specialists they could not help him, and having others tell him his nerves would never move his arm again, Rabe still hoped for some semblance of function to return. Hoping to find a surgical fix, he met with the foremost nerve surgeons in Washington State, trusting he might find an encouraging prognosis. "Everybody had told me, based on their training and expertise, I would never move my arm again." But Rabe wouldn't concede.

"Being in the medical field, I knew that the worst-case scenario is often communicated, but other options are there. I didn't, however, want to bounce around surgeons until I liked an opinion. I needed to find *the* guy or gal in the country. If he or she told me this was it, I could move on."

Rabe initiated a search for the best nerve repair surgeon in the United States. He believed he found that person in Dr. Scott Wolfe at the Hospital for Special Surgery in Manhattan. Using a procedure called "nerve transfer," Wolfe implanted new nerve tissue into Rabe's shoulder. If the surgery felt promising to Rabe, it was far from an immediate solution. "I had to spend hours and hours in the gym," he recalled. "Even when I'm home I'm staring at my hand, trying to get it to move. . . . I call it a Jedi mind trick because I have to just stare at my hand and tell it to move without being able to feel anything."

After six months of presurgery therapy, and nine months post, Rabe began "to get movement I did not have." Now he views his recovery as a ten-year campaign to achieve strength and return to working in a hospital emergency room. Fortunately he had good disability insurance and has been able to take on some teaching for a local paramedic program as well as occasional guest lectures with the University of Washington. "I tell my story quite a bit in my teaching," Rabe noted.

The year and a half away from working as a doctor hasn't been without rewards, Rabe said. He began coaching his son's soccer and baseball teams, and even traveled to out-of-town games—something he would not have been able to do if he was still working his emergency room shifts. Spending so much time with his sons, he said, "I felt like that was a gift."

There's also the aspect of setting an example for his children. "They got to see that life isn't always perfect," Rabe said. "You have to persevere. You get out what you put in." And then there is Rabe's own mental perspective on his life. "Some people don't get to experience the extremes of mortality, and it's been interesting to see the flows of joy, anger, and sadness." Rabe has rediscovered acceptance of life and its small, daily joys. "Even when I was just out of the hospital, I still felt happy with what I had. Don't get me wrong, I had days when I got angry. . . . I was short with my kids and with my wife." But his family learned to be part of his recovery. When he was short-tempered, his wife, Jill, would just say: "That's his arm talking, not Jake."

FOR SOME AVALANCHE SURVIVORS, THERE is a shift in their tolerance for risk, but not a complete abdication from it. Gary Ashurst, who went through a similar recovery from his avalanche injuries in France, concedes that he's now "a bit more conservative" in his skiing. But the mountains are where he found his life's inspiration, and he's not about to turn his back on them.

"I went forward with the thought 'I'm not going to live my life in fear,'" Ashurst said. And looking back on the accident that nearly killed him, Ashurst is sanguine. "If that's the way I'd gone out, then that's okay. It's more attractive to me than growing old and dying in a geriatric home."

Sam Kapacinskas, who survived an avalanche in Little Cottonwood Canyon's Birthday Chutes, and recounted his story to Craig Gordon's annual avalanche seminar, described the aftermath like recovering from an illness. "I think there's an element—at least for me—of being like a cancer survivor." Yet Kapacinskas hasn't abandoned forays into the backcountry, although his earliest returns were not without misgivings.

"I went on a tour almost two months to the day," he said, "and I was puckered the whole time." At one point, near the end of a backcountry run, he realized he could straight-line the rest of his ride down, and thought: "I'm not making any more turns. I'm getting the hell out of here." Kapacinskas looks at backcountry boarding through a different lens now, particularly when making decisions on when to drop into a steep pitch or not. "If I'm stepping onto a slope, I don't want to justify it just to me. I want to be able to justify it to my sixty-year-old parents in Florida. . . . I want to justify it to my life partner as well."

Like Rabe and Ashurst, Kapacinskas had a kind of awakening to the small, daily gifts life offers, something he's tried to share in talks he gives about his avalanche experience. "Even on my shittiest days, I can say 'I'm still alive.' I found a girl I'm going to marry. . . . It's been wonderful in the perspective it's given me."

26

THE TUNNEL CREEK AVALANCHE: MEGAN'S ECHOES

In twelve days Megan Michelson would mark the seventh anniversary of the Sunday when she stood among a dozen expert skiers on the crest of a run named Tunnel Creek in the backcountry just outside Washington State's Stevens Pass ski area. Seventeen inches of snow had fallen the night before. Beneath Michelson's feet was a snowy ledge leading to a steep pitch into trees, and beyond them a narrow chute choked with deep, fresh snow. From above, the skiers' anticipation was like that of a boat crew about to enter a tunnel of thrills. Nothing broke the spell of the moment. Nothing hinted at the horror about to engulf them.

"It shocks me that it's been seven years already," Michelson told me in February 2019. "I just have such vivid memories from that day. It could have been yesterday because I remember it so clearly. And I replay it in my mind still so much."

The day that will forever live with her was February 19, 2012. Michelson had driven to the ski resort that morning from Seattle with her fiancé, Dan Abrams. The couple had been invited by Stevens Pass marketing director Chris Rudolph to go on a backcountry tour for ski industry folks, and local Cascade Range luminaries, including Jim Jack, whom Michelson knew as "a legend at that mountain." Mostly though, the skiers had never met each other. As is often the case with a group of strangers, the gathering skiers sized up one another more with glances than words.

"Looking back, of course, with the hindsight of 20/20, I should have known there were some red flags," she recalls. For one, the snow fall had prompted a morning report cautioning of "considerable" avalanche danger, and several of the silent skiers knew about it, though no one shared their concerns. Moreover, they were headed into the out-of-bounds, beyond patrolled runs. "I was really deferring to the locals in terms of the judgment of that terrain," Michelson remembers. "I had skied back there once before, but they had skied back there countless times."

Michelson felt a persistent unease about the day's plan, but like others in the group, she didn't venture to speak. "It was hard to raise your hand and say 'Hey, maybe we should split up and make this smaller.' And I was just not going to overstep people who lived in the area, who were very experienced with that terrain. I just didn't have the courage to say 'Maybe we shouldn't be doing this.'" By the time Michelson and the others were perched above Tunnel Creek, there was a moment of reassurance as several skiers, including the marketing director, Rudolph, and Johnny Brenan, dropped into the run and disappeared without alarm. Then Jim Jack kicked over the edge.

"What Jim Jack triggered at the very top," Michelson says, "we didn't even see it rip under him." But immediately something looked wrong to Michelson and Abrams, who had hung back at the top from dropping in. An eerie cloud of snow rose from the canyon below, dusting treetops, as if a bomb had silently gone off down the cliff face. Had they listened closely, they might have heard the faint snaps of evergreen trunks splitting.

"As soon as he went over the blind rollover we were able to inch down and see there had been a fracture line and an avalanche had occurred," Michelson recalls, "but we had no idea of the magnitude of it. Because as soon as it started picking up speed it basically collected snow and it ripped . . . much, much wider. And then it traveled nearly a mile downhill. So it ended up being a massive avalanche that could have taken out cars it was so big."

Once Michelson and Abrams had edged their way into the top of the avalanche path, they saw it had scraped away so much snow there was only rock and ice. And no sign of Jim Jack. "I just instantly knew he had gone all the way down this huge ravine," Michelson remembers, "which was just a classic looking avalanche chute. . . . It was textbook. So I just knew instantly that he had been taken for a very long, very dangerous ride and that we had to get down to the bottom."

But side-stepping down was almost as threatening as the slide itself. "It was horrifying. Never knowing if we were under hang fire of another avalanche occurring." Michelson had the presence of mind to make a swift call to 911 fearing that Jim Jack wasn't alone in being caught by the slide. "Calling 911 was basically a back-up system. There was no chance that those ski patrollers were going to get there in time to help our friend. It was up to us. . . . We were calling 911 because we thought we might need a helicopter rescue if they were badly injured."

As she and Abrams descended, images of buried skiers crept into her mind. "We didn't know of the others who had already dropped in, if people had been out of the way of the path, or if they had gotten caught in it. We could have had up to seven people buried." Nearing the bottom of the chute, Michelson and Abrams came upon the wreckage of snow, rockfall, snapped trees, and the certainty that three skiers—Rudolph, Jack, and Brenan—had vanished. One skier, Elyse Saugstad, who wore an inflatable avalanche rescue backpack, was partially buried but mostly unhurt. The backpack had buoyed her face just above the snow.

Michelson, Abrams, and others began a frantic beacon search for the signals of the three men's own beacons. "I had my beacon on, I was ready," Michelson says. "We needed somebody to direct and I got there, one of the first people to get to the bottom where I started getting signals. So I just started doing what you had to do. It wasn't because I was braver than anybody else, it was just because I was there and there was no choice."

As Michelson homed in on beacon pulses emitted from beneath the snow, other skiers began digging. "The three men who were buried were very deep down in very heavy snow, so it was a laborious process getting them out." Scared as she might have been, Michelson knew she had to wall off those feelings and channel all her energy on the rescue. "I feel like you don't really know how you're going to act in an emergency situation unless you've been in them before and until you're in them. I am proud of, I think, the way I tried to keep myself together knowing that the situation demanded that kind of focus. There was no time to have an emotional reaction to what we were going through. You just knew that your friend's life was at risk and that they were counting on you to get to them."

Uncovering the men ate up minutes their lungs and hearts couldn't withstand. Despite efforts to revive them, Rudolph, Jack, and Brenan, a father of two, were all dead.

IF THE AVALANCHE FELT LIKE what Michelson understandably calls "the worst day of my life," the media aftershock that followed was almost equally wrenching.

At the time, Michelson worked as an online freeskiing editor for ESPN. "As soon as the avalanche happened, I just had this pit in my stomach because I knew we were going to have to cover this. Three people dying in an avalanche is news alone, but having those guys who were ski industry people, I just knew this was something I was going to have to cover in some way, which is a horrible feeling when you are trying to process the grief of what just happened to you."

She anticipated an onslaught. "I understand why people want information at that time because people do need information in order to figure out what's happened and obviously we wanted to get the right information out there. But when you are in it you just wish that people could lay off for forty-eight hours so that you can just try to process what's happened and have a moment to yourself."

But the media world didn't give Michelson and Saugstad, who had gone to sleep at Michelson's Seattle home, even a single night's rest. "We had producers [calling] from *Good Morning America* and the *Today Show*. I went with Elyse to like a 3:00 a.m. shoot, morning television shoots that following morning after the avalanche, basically the middle of the night."

After the national television shows and newspaper reporters had moved on to the next news cycle, *Outside* magazine commissioned Michelson to write an account called "Tunnel Vision." Memories of the day were still emotionally raw. "You have survivor's guilt, there's so many emotions. You have post-traumatic stress disorder, potentially. You have all these things you are going through."

Just when months had passed and it seemed much of the world might have forgotten the story of Tunnel Creek, Michelson and other survivors were approached by John Branch, a reporter for the *New York Times*. He wanted to write a lengthy, hyper-detailed story about what had gone wrong when such an experienced group of skiers, trained in avalanche awareness, had seemingly ignored numerous signs of danger in the backcountry. The front-page story was called "Snow Fall" and Branch received a Pulitzer Prize for it.

"When John Branch approached us . . . at that point I was actually ready" to talk about the accident, Michelson says. "I knew there was a benefit to getting our story out there." The story received such wide readership and interest that it eventually became a case study in avalanche education courses. "For me it's bittersweet, but I am glad that people can learn from the mistakes we made, and use our group dynamics, and the human factors that went awry in our situation, as a learning tool."

MICHELSON AND ABRAMS WERE MARRIED a month after the accident, and eventually they moved to Tahoe City, Nevada. But in the weeks and months after that day in February, life was anything but evenhanded. Before their wedding, the couple felt compelled to meet with a trauma therapist.

"We saw a therapist because it seemed like the preventative, right thing to do," Michelson says. "It was just so much stuff, so many emotional things, highs and lows." More lows than highs afflicted her for a long period, and even after seven years her memories and emotions about the day remain delicate. "The nightmares and lack of sleep and reliving it, over and over and over again in your head, every time you had a quiet moment, that happened for months. And I still sometimes, I'll be out doing something and I'll have a moment to myself to think and I just instantly get flashbacks of that moment. Maybe it's just your body's way of still processing it, or, I don't know why we do that to ourselves? But I still do it. I don't think you ever, ever recover from seeing that, and going through that, and losing friends."

Whenever avalanche accidents flash to her attention, there's an emotional jolt, a surge of memory mixed with empathy. When the October 2017 Imp Peak avalanche in Montana's Madison Range killed Inge Perkins and led Hayden Kennedy to take his life, the tragedy dredged up nightmarish feelings. "I remember feeling the shock hearing about that, but also completely understanding. He probably came away from that . . . and not being able to save her, or to prevent that from happening, he probably just. . . . It sounds insane, but I think that would be a really, really hard thing to live with your significant other dying in an avalanche."

Sometimes, even without the emotional cut of an avalanche report, Michelson suddenly experiences a welling up of feelings, echoes of phantom-like details from Tunnel Creek. She knows she can call on her fellow survivors for reassurance. "Every once in a while we reach out if you're out having a hard day, or you're thinking about that day. And you know you always have those people you can connect with and they will understand. Even now, seven years later, they understand if you shoot them a note and say, 'I'm thinking about that day and just want to give you a virtual hug.' They understand."

Today, Michelson is the mother of two young children who are beginning skiers. After first thinking she might never again return to the backcountry, she finally did but with a completely new set of personal rules. "I have changed my entire approach to the days that I ski, who I ski with, how I approach the backcountry. I'm much more cautious now. And I also make decisions for myself now. And I would never go in a group setting like that where I wasn't the one choosing what terrain I was going to ski."

Her determination to ski safely is hard-won, and while she doesn't think every skier or snowboarder learns only through tragedy, she believes some people are too easily seduced by a slope of untracked powder. "I don't think anyone looks at a run and says, 'This is worth dying for.' No one thinks that way. But I do think people look and say, 'Well, the chances of an avalanche happening here are low and I want to ski some powder.' There's all these things: 'It hasn't snowed in a while, and I've got this desire to get out there. I'm with all these friends and I haven't skied in a while. Conditions are perfect. I've got the day off from work.' All these different factors that we just, standing on top of runs, [weigh] making the decision to ski it. I don't think anyone would ever say, 'This is worth orphaning my children over.' But I think people trick themselves into thinking it's going to be safe and they're going to be fine. And we obviously did that day."

As she approaches another anniversary of the accident, Michelson circles back to the bond she shares with the other survivors. "Every year, on that date, I try to do something in memory of the three men that we lost that day. And I also usually reach out to the other people that survived. It's a sad day, but it's also a day to celebrate those men and the fact that we're still here. . . . It was the worst day of my life, but it did bring me closer to the people I was with that day."

The path forward might have been easier for Michelson if she declined to talk about Tunnel Creek. Seven years is a long time to confess mistakes. But she believes that the more backcountry training people get, the more they understand what causes things to go wrong, then the better chance backcountry adventurers have to stay alive. She's willing to help people stave off the white dragon.

"I think it's just really important to get that education out there so that people understand the risks they're taking and know what steps they need to take before they head out-of-bounds. . . . And if a tragic story like the season of the Tunnel Creek avalanche can help remind people of that, then at least we didn't lose those three men for nothing. At least we can use their stories to help other people make better decisions than we did. That's really the only reason why I'm still talking about it."

27

A CRYSTAL BALL'S FUTURE TROUBLE WITH CRYSTALS

If the winter of 2017–18 ended with an underwhelming, sparse snowpack, the northern Utah summer that followed felt as parched as the dusty rock paving a dry Mexican riverbed. Out behind my house one August afternoon, sitting beside the driftwood dry roots of cottonwoods, I glanced up to see tendrils of brown smoke rising from the ridgeline bordering Powder Mountain ski area. For the next several days, firefighting airplanes made constant, daytime touch-and-go maneuvers over nearby Pineview Reservoir, filling their bellies with water, then bombing the flames below in brief waves like synthetic cloudbursts.

Started by stray bullets on a nearby gun range, the fires were eventually smothered, but even after the flames and smoke were gone, the air over the Ogden Valley lacked any trace of humidity. The Salt Lake Valley smoldered at the foot of the mountains where signs shaped like Smokey Bear reported daily that fire dangers were "high." To the east, the city of Denver, a news report proclaimed, "was drier than Phoenix in 2018." The city saw only 60 percent of its average annual precipitation. Following the 2016–17 winter that marked the second lowest snowfall in Denver's recorded history, the 2017–18 winter had improved only marginally—the fifth lowest snow total on record. When early snows arrived in the fall of 2018, painting white the Colorado Rockies and Utah's mountain ranges and those in its neighbor states, people—especially skiers and snowboarders—celebrated the bountiful return of winters past. Maybe the climate change outcry had been akin to a meteorological boy repeatedly fantasizing a prowling wolf.

It was even a little hard to reconcile climate change predictions with the November and December snowstorms that barreled east from the Sierra into the Wasatch and continued to build a snowpack that, by mid-February 2019, was at 123 percent of the season average. Seeking insight, I drove down to Salt Lake City to see Brian McInerney at the National Weather Service headquarters. Technically, McInerney is one of the Salt Lake weather service's lead hydrologists, a true water scientist who started working for the service in 1990 and never left a "great gig" as he calls it because he can practice weather science in a place that also offers some of the best powder skiing on the planet.

Officially, he is also, in federal government speak, the "point of contact for climate change information for the forecast office." But McInerney, whose soberingly serious demeanor can effortlessly shift to an easygoing philosophical lilt, describes himself simply as the Salt Lake service's "climate change guy." Or, when in a pessimistic mood, he will say of his climate change duties: "I do this for a living. I just depress everybody I talk to. Like that Debbie Downer person on *Saturday Night Live*." McInerney has mastered the increasingly delicate talent of sharing data-driven scientific insight into climate change with the more indelicate message that, if emergency actions aren't taken, skiing in the Wasatch—and some world civilizations as we know them—will cease to exist by the end of the century.

One of my first questions to McInerney was how one could square the alarmist sentiments about climate change with the exceptional powder skiing I'd been having on a fat winter's snowpack. "I think the naysayers get weather and climate confused," he cautioned. "Weather," in other words, was last week's fresh fourteen inches of powder snow. "Climate" is the trend in precipitation and temperature over the past thirty years and the next thirty.

The same morning I met with McInerney, a report was released that described how two climate scientists had designed a computer model projecting out to 2080 the likely effects of an altered climate on any given American city. The model showed that the "new" climate for a city would be roughly analogous to that of an existing city 528 miles south in latitude of it. For example, Washington, DC, would feel more like southern Arkansas. Los Angeles would experience the climate of Las Palmas, Mexico, on the thousand-mile-long Baja Peninsula. And in Northern Utah? McInerney says that by 2100 there is the potential for Salt Lake City's average climate to warm by ten to twelve degrees. "So the climate of Salt Lake City will become the [new] climate of Park City," he says. Only about a thirty-minute drive into the mountains from downtown Salt

Lake, Park City has an elevation of 7,000 feet—nearly 3,000 feet higher than Salt Lake's 4,226 feet.

Rising temperatures are only one part of what McInerney sees. The annual profile of hydrology is also changing, particularly as it concerns snow. "It's still predominantly snow," he says, "but that's changing and it's changing pretty quickly. What we're seeing now is more rain in October and November than we used to. And then the winter season is ending sooner with high pressure and it's becoming warmer and drier. And we've been melting prematurely."

When early melts happen, McInerney says, the runoff is "less efficient. You have less water coming out of the mountains." In past years, people would ask McInerney if varying snowfalls and rain were due to "natural variability or climate change." He explains: "Twenty years ago you'd be like 'I don't know, I'm not really sure.' And then, the more we went into this, the more you could see just how out of whack temperatures were when we would just keep breaking heat records, and you would see more rain, and you would see high pressure dominate the weather during the wintertime which makes for a weaker snowpack." Concerns about that weakness are what keep McInerney and avalanche forecaster Craig Gordon and his fellow forecasters awake at night. "You see the danger of the avalanche conditions becoming increased because you don't have as many storms and you are getting rotting snow," McInerney says. "That's something we see more of. And then you say 'Well, this is going to reach some magical time and we'll just go back to the way it was.' And unfortunately it's not."

The winter of 2018–19 delivered far more snow than the prior season, but with it came a sort of reverse danger than with thin, decomposing snow. In Colorado the seemingly endless string of snowstorms—many of which dropped heavy snow high in water content—spawned a torrent of avalanches. By early April the Colorado Avalanche Information Center (CAIC) documented that forty people had been caught in avalanches, fifteen of them buried, and eight killed. The snowpack reflected attributes, like dry spells followed by intense storms, that McInerney cites as troubling signs of climate change.

"We had early-season snow—snowfall in October and November that didn't melt away," Brian Lazar, the CAIC deputy director told the Denver-based publication *Westword*. "It sat there long enough under some early-season dry conditions to turn into a weak layer of snow, which made for a poor foundation for the snowpack." Then a series of big, wet storms hit the Rockies. "That built thick layers of snow on top of this weak foundation," Lazar explained. "And when you're loading weak layers at the bottom of the snowpack with heavy snow on top, that's conducive for creating avalanches."

These reports have hardly been limited to the American West and the Rockies. An August 2018 article in *The Guardian* detailed the dire retreat of glaciers in the Alps. "In the Alps, the glacier surfaces have shrunk by half between 1900 and 2012 with a strong acceleration of the melting processes since the 1980s," said Jacques Mourey, a mountaineer and researcher delving into the role of climate change, particularly in the peaks above France's Chamonix. "A 1970s climbing and mountaineering guidebook to the 100 best routes around Mont Blanc isn't usable any more," he told *The Guardian*, "as most of the routes have changed and can't be used. . . . If anyone doesn't believe that climate change exists, they should come to Chamonix to see it for themselves."

Eight months later, in April 2019, *The Guardian* ran a story on a just issued study by Environment and Climate Change Canada, showing that nation's climate is warming twice as fast as the rest of the world. It noted that while global temperatures have increased 0.80 degrees Celsius since 1948, Canada's have spiked 1.7 degrees Celsius. Researchers attributed this to the burning of fossil fuels. Some snowpack forecasters, like McInerney, have indicated that as less snow falls in Utah's mountains and melts under warming winter rains, the snowline for skiers will advance farther north into Canada. But the study "paints a grim picture of Canada's future, in which deadly heat waves and heavy rainstorms" reshape the climate much like Australia experienced in the summer of 2019–20. Needless to say, if such a scenario plays out, skiers going north for snow would find even scarcer snowpack—if any.

At the forefront of lobbying on behalf of skiers, snowboarders and other snow sports enthusiasts concerned about snow devastation is Protect Our Winters (POW). The Boulder, Colorado–based group is made up of a dozen staffers, 150 elite athlete ambassadors, and about 6,000 paying members. Through state and federal lobbying efforts, political activism, and other actions, POW seeks to preserve winter as a field of play for skiers and snowboarders as well as for the outdoor sports companies, ski resorts, and snow professionals who constitute 7.6 million jobs and $887 billion in annual revenue. Those numbers, POW has documented in its economic studies, don't include the twenty million people who annually ski, snowboard, and snowmobile in America's mountains. Through lift tickets, resort hotel stays, and restaurants, those people spend an estimated $20.3 billion, says POW. "There's a real economic argument here," says Mario Molina, POW's executive director.

POW was formed in 2007 by professional snowboarder Jeremy Jones of Truckee, California, who remains an ardent board member. "He's very actively involved and very savvy about helping get us to where we need to be," Molina says of Jones. "Having him as a founder has given us access to [elite] athletes. There are a lot of athletes who look to Jeremy as a role model."

Situated at the center of the snow dissipation threat are the ski resort companies that have invested hundreds of millions of dollars in infrastructure—real estate, ski lifts, enormous base and mountainside lodges. Not to mention lots of operations people—executives, managers, ski patrollers, instructors, "lifties," and other service workers. Slowly, Molina says, the resorts are beginning to lobby around climate change concerns. But because many operate on federal Forest Service–controlled land, and climate change is a politically charged issue, they are reluctant to shake the trees too hard.

Aspen Ski Company has assumed a strong advocacy voice, says Molina. "They've gone out on a limb and made this a priority. They're taking a risk that they are going to be on the right side of history." Molina and POW are trying to coax other companies that have undertaken solar and other environmental energy projects, but political activism hasn't been a natural, historical identity for these companies. "The industry has been slow to understand," he says. "It's fantastic if they green their operation. . . . Those are all great things they should be doing." If "the entire ski industry could reduce their footprint to zero," says Molina, it would mainly serve as a mobilizing symbol for America's twenty-million-plus snow sports enthusiasts. He sees POW's role as just one among thousands that have about twelve years to make critical "green" initiatives if climate change is going to be materially slowed and rising mercury gauges arrested.

Sometimes snowy winters like that of 2018–19, as much as they provide a memorable season, they don't necessarily make Molina's job easier. "This winter," he says, "people say half jokingly 'We don't need you.'"

MOLINA IS FOCUSED ON THE next twelve to thirty years, but like all meteorologists, McInerney concentrates his energy on predicting what the weather will do five or so days out; that's what most people are looking for. Often he's part of a National Weather Service–synchronized launching of a daily 4:00 p.m. weather balloon.

Tethered to a radiosonde, the balloon sails skyward until it's at about 110,000 feet, feathering space, where the sensors send back weather data used in assembling forecasts across the country. The extreme low pressure at altitude causes the balloon to pop, and the radiosonde descends from what McInerney calls "a GI Joe parachute." If the image of what seems an almost anachronistic practice—flying a balloon to record the weather—occupies some of McInerney's time, his focus is increasingly on computer predictions that go out nearly a century.

The timeline he's mapped is beyond shocking. For skiers and snowboarders, whether in resorts or the backcountry, it could lead to the global transformation (and ultimate end) of a sport and way of life. Between now and then, the likelihood of more dangerous avalanches will announce themselves in more of a bang than a whimper. "You start putting the pieces together and you learn that by 2035 to 2065, areas that are 100 percent snow covered in the Wasatch are only going to be 50 percent snow covered, or less. . . . What that means is the base areas of these ski resorts at 7,000 feet or 7,500 feet aren't going to have snow. And so the financial model for the ski areas, if you can't have snow at the base, what are you going to do? Put people on gondolas and haul them up the mountain? And after a while that snow line is just going to increase [higher] until you hit 2070, 2080, where you won't really have snow on the mountains."

As McInerney describes it: "Probably, by 2100, you'll only have snow in December, January, and February in Big and Little Cottonwood canyons, at the top and the Uintas, and the rest of the Wasatch is going to be snow-free . . . . We won't have snow. We'll just have rain during the winter time. That's a tough one to take on so many levels." The fallout in this timeline isn't limited to skiers and the recreational businesses that serve them; after all, it's a lifestyle and economic ecology formed around fitness, elation in the mountains, and ultimately fun. Where the most serious damage will be is in shrinking water supplies for farmers and communities.

McInerney conjures an image of the perfect system that has served the American West for centuries. "We start snowing in October, we store it, and then when the spring weather comes around, a cold, wet spring stores it as long as possible. And then we flip the switch, turn the sun on. It melts, it's clean, it's gravity fed, it comes down channels that have been formed for thousands of years into the reservoirs. And then we dole it out in the summer months. . . . What's going to happen when we have rain during the wintertime, when we don't have snow? And it comes in these intense rain events that we can't capture. It just runs off. What are we going to do for water supply? Let alone powder skiing is going to be gone. . . . The canary in the coal mine will be powder skiing. . . . It just portends our future for our kids and our grandkids."

I told McInerney about my two-year-old grandson, Clark Hill Power, who lives in Maine. My aspiration to teach him to ski in a year or so will likely happen. But after creating all that winter wonder and happiness in his life, I'll one day pass away and leave him to wonder why my generation and preceding ones had not acted responsibly to preserve this shared joy of sliding down snowy slopes in the mountains. Gone might be an inheritance of nature's joy for his generation and others to come.

"When your grandson, when he's an adult," McInerney says, "he'll look at our legacy and say, 'Why did they wait? Why did they not do anything about this? Why are we now having to deal with their inaction and their indifference and apathy?' That, I think, is a pretty tough legacy to have for us. We could have done something."

28

LATE SUNSETS: THE BACKCOUNTRY IN SUMMER

In summer, high in the mountains of northern Utah, the sun never sets before 9:00 p.m. Above about seven thousand feet, the melted snow of ski runs gives way to thistles, fireweed, sunflowers, and other wildflowers that mingle together and hide rattlesnakes among their roots as if the snakes were the nest-fleeing offspring of much larger reptiles disappeared until a new winter arrives. Hanging over the summer slopes is the smell of high desert—desiccated sage and soil mixed with rich clay. It's as if the air is wafting from the lip of an aromatherapy bottle. In some ways, it's a time of hibernation for all involved in snow and the mountains. Season tallies have taken place—inches of snowfall, numbers of skiers on the slopes, injuries recorded, avalanches encountered. And deaths investigated.

For the snow patrollers, avalanche forecasters, backcountry guides, and others, seasonal jobs for the summer take hold. They create weddings in the mountains, shoot photographs, put on craft beer tastings, schedule live music on stages beside ski lodges. Others disappear into the mountains, or onto the water, rock-climbing, kayaking, rafting, guiding in heat and sunlight, and on river rapids, just as they led backcountry riders into the cold and snow.

And they wait. Winter—and its white dragons—is never far away.

DOUG CHABOT USUALLY LEAVES BOZEMAN for the summer to climb mountains in places like Pakistan and Switzerland. He also helps at a network of schools he cofounded in 2011 to teach Pakistani girls to read. Under the auspices of the Iqra Fund (*iqra* means "read" in Arabic), the program has grown to oversee seventy teachers and nearly four thousand students, a share of whom are now boys. Chabot launched the program with Genevieve Walsh, a Bozeman resident to whom Chabot was once married. Walsh wrote a doctoral dissertation on education in Pakistan and out of that work the Iqra Fund was born. The primary school track goes from first through seventh grade.

"She's the brains behind it," Chabot said of Walsh, communicating in July 2018 on Skype from Afghanistan. "I got to say, it's the coolest thing I do." In Afghanistan, Chabot worked with the Agency For Habitat Disaster Relief to help train residents of mountain villages to identify avalanche paths and plan for slides that have had a history of destroying tiny villages and killing people. With his winters filled by avalanche forecasts and frequent reconnoiters of the backcountry snowpack, Chabot relishes his summer travels in such foreign mountain ranges as those in Afghanistan and Pakistan. Time away usually helps him forget the stresses of winter.

The winter of 2017–18 had begun tragically for him, when Inge Perkins and Hayden Kennedy died, and it did not offer an easy ending. On Saturday, April 14, about midmorning, a lone skier left the boundaries of the Bridger Bowl ski area and slipped into side-country terrain. He hiked to the north summit of 9,100-foot Saddle Peak. Just before 11:00 a.m., he dropped into a run to the east, and an avalanche broke. The skier was carried 1,500 feet down the mountainside and buried.

A mile from Saddle Peak, a skier riding a chairlift saw the avalanche suck the man under, and quickly alerted patrollers, three of whom swiftly made their way to the avalanche field. Using search beacons, they found a signal and eventually a gloved hand reaching out just above the snow. Nearly seventy-five minutes had passed. Pronounced dead, the skier was an official fatality by avalanche. Chabot was called in to handle the investigation and a helicopter flew him to the site. Halfway through April, at what he hoped would be an anticlimactic end to winter, Chabot was worn thin from seeing death in the mountains.

"I have to say, here it was mid-April and I was so weary after such a long season. It was a feeling I'd never had before. I was mentally tired." Escaping to Red Rocks in the desert outside Las Vegas, Chabot and his girlfriend got in some climbing, the terrain and warmth a strong antidote to "a long season of being mentally taxed."

Then he headed to Afghanistan for his avalanche awareness work, which bore its own variety of risks and stresses. To make himself less distinct,

Chabot grew a beard and adopted the local dress of the Afghani mountain people. Unlike some of his previous trips to the country, however, he sensed an increased threat. "The Taliban have been making some pretty dynamic strides in northern Afghanistan," he said. With the Taliban's increasing presence, even the mindset of the villagers he encountered had shifted. "They've settled into the fact that it's going to be pretty chaotic," he said. "It's sad. But it's reality. They're looking at some hard years ahead of them."

Still, even over the Skype connection, Chabot's voice was buoyed by a lilt of optimism, born of his hope that his work would make a difference and by the prospect of more climbing adventures, one of which he'd soon embark on in Switzerland. "We hope to climb the Eiger, if the weather lets us."

In August he would fly to Pakistan to spend more time with teachers and students involved in the Iqra Fund. "Getting a girl to go through seventh grade is life changing," he said. "She stands a much greater chance of not dying in childbirth. If a mom has been educated, the odds her kids will get educated go way up." Several hundred of the Iqra Fund students have gone on to secondary schools in larger cities, Chabot said, and even come back to their villages as teachers. "It's like magic. It's unbelievable."

He knows Bozeman and a new winter await him. But for now, his backcountry is in other countries. "It's a summer away just enjoying the world, a little play and a little good work. It's a time to recharge." Like his avalanche awareness work in the Gallatin Range, Chabot finds bonds between that and his summer endeavors.

"You just have to want to do a little good in the world."

ANDREW MCLEAN LIKES TO PEDAL up mountain trails above Park City in the summer, riding not a knobby-tire mountain bike but a unicycle. "You get a really good workout," he said. "After an hour you're pretty cooked." Because unicycling requires the rider to maintain "a still upper body, it's similar to skiing."

It's just one of the ways McLean conditions himself for the trips he usually takes in the months after Utah's winter ends. Which doesn't mean he isn't parachuting down into winter somewhere else on the planet. His plan in May had been to do a skiing trip to Greenland. But a lack of snow there forced a change in geography. "They had no snow, which is scary given that they're around the Arctic circle." Another possible signpost of the changing climate. So McLean and a companion made their way to Switzerland, where the winter had produced a deep snowpack in many places. McLean skied in Chamonix, where he'd surprisingly never been before, and because it's so easy to travel around the Alps, he was able to visit a number of friends. "It was fun to see and ski with such a wide variety of old and new friends," he wrote in a blog about the trip. "Skiing is all about being in the moment, but it is the experiences with friends and shared experiences that make it so memorable."

His introduction to Chamonix began not on the snow, which he said was "so bad it was almost unskiable," but rather in the air. A friend, Ruedi Homberger, took McLean on an early morning flight over Chamonix and a number of the most famous peaks in mountaineering history. Homberger "gave me the sunrise grand tour of the area, complete with names and notable ascents/descents. After reading about these landmarks for years, it was surreal to finally see them all—Mont Blanc, the Aiguille du Midi, Walker Spur, Grandes Jorasses, Heilbronner, and the Gervasutti all lit up in morning alpenglow."

After Chamonix, McLean went to Verbier and did some backcountry skiing with friends for a day. They headed up to the Great Saint Bernard Hospice, a historic overnight cabin that dates back to 820 AD. It also is the origin of Saint Bernard dogs used by mountain guides and sometimes in rescues. "The Saint Bernard hut has been destroyed or burned and rebuilt many times over the centuries," McLean wrote in his blog, "and I'm guessing it currently can hold 100-plus people during the summer. . . . While we were there, it had about 15–20 guests, most of whom were passing through as part of the Haute Route."

Back in Utah in July, McLean was preparing to depart again so he could visit his wife, Polly Sammuels, a lawyer and champion backcountry skier. She was in Courcheval, France, along with their two daughters, taking part in a job exchange program as part of a sister city relationship that Park City has with Courcheval. When McLean got to France, the family would be taking a Via Ferrata tour into the Italian Alps. In English the words mean "iron roads," describing a series of consecutive climbs enabled by iron rods, cables, and handholds embedded into the rock since World War I. The holds were originally constructed so soldiers could efficiently get themselves and supplies into the Dolomites, where they were waging war. McLean describes it as "a good family activity"—his way of saying it's pretty challenging and requires a high level of fitness; it's not for those spooked by heights.

Now fifty-seven, McLean has seen, through many years in the mountains, a number of his friends die in the backcountry. His knowledge, experience, concern for safety, and willingness to bear up in the worst of outdoor elements have made him a sort of elder statesman of the backcountry. But another trait has served him just as well—his sense of humor.

Of his time in Chamonix, McLean wrote: "It is an area with instant and easy access to radical terrain and no shortage of testosterone to temper it. In this regard, it reminds me of renting a piece of heavy earth-moving equipment with zero training on how to use it—you can quickly get yourself into trouble."

Even bedding down for the night on a ski trip can prompt McLean to find humor in what otherwise might be an annoyance. "We found a last-second room at the historic Hotel Gustavia," he wrote, "which turned out to be a screaming deal in part because it was located right on top of three very popular bars."

SHANNON FINCH TRAVELS INTO A cold, crystalline backcountry in summer, though instead of skis and poles she employs a kayak and paddle. It's the summer backcountry setting where she began developing her "guide voice." And it's a place where, on some of the country's best whitewater, she returns for the rhythms of the river, just as she returns to the mountains for the physical synchrony of turns in powder snow.

After backcountry guiding ended for the season, Finch drove her truck, loaded with kayaking gear, to California's Kern River near Bakersfield. She describes the Kern as a "wild and scenic dam free" sweep of water, meandering its way down the gravity path of snow melting under the Sierra sun. But a river like the Kern feeds on snowpack, and because the winter had produced a lower snowfall, Finch only spent about a month teaching kayakers and making river runs on her time off.

She then headed to Buena Vista, Colorado, to kayak on the Arkansas River and do some branding work for Kokotat, a manufacturer of technical gear for kayaking. "The Arkansas offers Class 2 to 5 whitewater," she said. "I rendezvous there with friends from the river community." In early June, representing Kokotat, she headed to Vail for the GoPro Games. "It's a big mountain festival that anyone can participate in—mountain biking, rock-climbing, slack lining." Usually sleeping in the rear cab of her pick-up truck, often with her search dog, Leif, tucked by her side, Finch made her way in mid-June to the North Fork kayak championships on Idaho's Payette River. It's an invitation-only competition for "elite" kayakers. Although Finch does compete in kayaking challenges, she was at the North Fork on behalf of Kokotat.

Listening to Finch describe her summer peregrinations and understanding the energy it all takes can wear one out—and she was only through June. On the threshold of July, she drove back to Utah, where she took a paragliding class. "I'm twelve days into the class now," she said. "Anywhere from ten days to twenty is the window. It's weather dependent."

Every morning since the class started, she headed to Point of Mountain in Lehi, Utah, where from 6:00 a.m. to 10:00 a.m., she learned the fine points of parasail wing trim and weight distribution, feeling her way among the wind currents the same way she feels her way through deep snow in the winter. "My ultimate goal," Finch said, "is to become a mountain pilot, so I can hike around the Wasatch and fly." These summer backcountry exploits have similarities to her winter ones. "The most similarities are in choosing your partners," she said. "Running whitewater, I have similar partners to those I'd choose in the [winter] backcountry."

After being awarded her parasailing certifications, Finch planned to head north to compete in late July in a kayaking event, the Toby Creek Race, outside Calgary in Canada. Registered for the elite division, she expected to be on Class 4 whitewater and would make timed paddle runs through slalom gates, both those going down the river and those requiring her to reverse paddle direction for upriver gates. "It's for money," she explained, "but it's also for fun, something to set your sights on and see how you're doing in the river community."

Given two months of such physical exertion, and some summer search training with Leif, Finch said August would be a time for vacation. Planning to head up into the Pacific Northwest, spots in Oregon and Northern California, she expected to do more kayaking and throw in some mountain biking.

And there was always a chance for pulling out her parasail. "My goal is to be a conservative mountain pilot. I am definitely playing it safe. I'm not looking for another adrenaline sport. I just want to have fun."

TITUS CASE MIGHT HAVE BEEN sharing some whitewater kayak runs in one of the same rivers as Shannon Finch, but his sixty-seven-year-old shoulders had other ideas—like surgery. At least that was what he was contemplating in early July. "My shoulders are just toast," he lamented, but he was still weighing the unknowns of surgery.

What must have left him struggling were his memories of the water. The Lochsa River. Glass clear and dusted white in places by the frying pan shoals where rocks congregated, clamoring in the current as they rubbed together their ancient stone casings. Flowing from north-central Idaho, the Lochsa's waters are fed by snowmelt in the Bitterroots, the water running south seventy miles, the water descending two thousand feet like a liquid couloir, the water marrying with the Selway River, and the two great arteries feeding the Clearwater River in the Nez Perce Clearwater National Forest.

Lochsa. Nez Perce for "rough water." Case had kayaked these waters for decades. His shoulders had wrestled with the water, his boat bobbing through the rapids like a bird diving and climbing between the whitewater waves. "It's a great, great river," Case declared. "It runs from the Idaho/Montana border down the drainage, west of Missoula, east of Lewiston." His voice imparts devotion, the romance received from a lifetime of river running. The Lewis and Clark expedition had wanted to run it, Case said, but after studying it, they concluded it was too rough.

The water summoned him to other places. "In my younger days, I did thirteen different trips down the Grand Canyon." Now Case is getting to know summer in the Salt Lake Valley a little more intimately. "I've been doing my usual hiking and biking, and playing golf. I try to mix it up. One day I hike, the next I bike, and the next I play." He is drawn to hiking in the backcountry near Alta ski area, where the terrain is inscribed in his memory. "It's a whole different thing to see the terrain in summer. It can be very revealing. . . . You have an appreciation for the terrain features that get buried in the snow." If Case courted solitude on winter mornings riding Alta's chairlifts to assess the snowpack, he says his habit of waking early reaches into summer. "A lot of my summertime activity, I do alone. To beat the heat and the crowds I get out early. In the Wasatch and in summer it can be a zoo."

Over the years, as more skis poles are swapped for hiking poles, or kayak paddles, Case has seen the crowds swell. "We're sort of loving the Wasatch to death. But it's great to see so many people out there." In August he'll return to his hometown of Rochester, New York. "My mother is having her ninetieth birthday." Though he is officially retired from his forty-two years at Alta, he hopes the coming winter will have him filling in again on avalanche work at Alta. The snows, like the waters of the Lochsa, have an eternal hold on him.

But before Rochester and winter, Case may go up to McCall, Idaho, to visit friends. Not far from McCall are the waters of the Lochsa River. For Case, it was not just life in the backcountry of summer, or a wait for the backcountry in winter. Looking back, he understands he's now in the astonishing backcountry of his life.

STEPHAN DRAKE, STANDING IN THE early morning glow above the snow on Mount Hood, Oregon, knows this isn't exactly the backcountry, but it's a perfect stand-in when you are looking to test skis in the United States—in July. "It's nice," Drake says of Mount Hood, "because in the summer, if you're a passionate skier or ski designer, you're heading to South America or New Zealand." But having taken such trips to winter climes, when it's summer here, involves significant expense and time, Drake knows from experience. To save on both, he can load thirty or forty pairs of skis from his company, DPS, into the back of a truck and in ten hours he'll be making turns on Mount Hood.

Drake actually has a lifelong love affair with the mountain. It dates back to his childhood when he came to Mount Hood in the summer to slake his youthful craving for skiing until winter arrived. He still lodges at the Huckleberry Inn, a place he stayed when a boy, there in Government Camp, the small town that serves as a kind of barracks to Mount Hood. Of the Huckleberry Inn, Drake says, "It's a little ratty, but pretty nostalgic and fun." In its rowdiest days, the town could have been the movie set for a sheriff and bandit showdown. Now, mountain climbers and skiers share time on pool tables, drink beer, and talk about the weather.

Mount Hood's ski runs are on a glacier, which helps preserve the snow long past when most ski area runs have melted to mud. The snow is treated to make it fast under your skis. "It's not totally ideal," Drake says, "because you're dealing with limited terrain and salted snow. . . . The snowfield up here is maintained and prepared for ski racers." Drake and his fellow testers are on the snow by 7:00 a.m. while the sun is still low. As the temperature rises, and the July sun starts to cook the snow, it becomes slush by about 1:00 p.m.

For the roughly five-day testing session, Drake brought an array of skis, some "fully conceptual," essentially experimental, and others that are planned for release. After that, the test skis will be shipped to sales reps for use in marketing. He also brought along two material scientists to test skis that had had their bases coated with Phantom—an exotic chemical developed by Drake's company that eliminates the need to wax ski bottoms. He may test some limited-edition skis designed and built under the name Powderworks, an exclusive arm of DPS that works with his best clients. These skis, tailored to each customer, are sold directly to clients looking for a ski that performs best in only a certain kind of snow or skier access such as heli-skiing. "We do these exotic, surfy pow shapes," Drake explains of the designs.

One thing missing from the test session, Drake says, is his DPS ski boot project—referred to as "Das Boot"—which has been under way for more than a couple years. "I'm lamenting I'm not testing it right now," he says. If there is any complaint that is almost universal among skiers, it's dealing with ill-fitting boots that pinch toes, bruise shins, and generally make one's feet feel like they're squeezed into a vice. Custom boot liners have helped many skiers, but some boots still have a "turn of the screw" temperament to them. Drake's mission is to solve all this discomfort. "We're shooting for the end of 2020 to first reveal it," he says.

Back in Salt Lake, Drake will spend some of his summer in the Wasatch riding his mountain bike and also pursuing a new hobby—racing sports cars. At the Utah Motorsports Park in Tooele, he turns laps in a 2001 Porsche 996 Turbo. "I meet skiers there all the time. Everyone gets the similar sensations of G Forces."

Ski testing on Mount Hood and car racing near his home base of Salt Lake are ways Drake has conspired to get work done and have some fun. But sometimes he longs for his younger days. "In my case it's probably a story of becoming more involved in the pressures of the business side," he says of his obligations running DPS. "In my twenties, I'd pack up and head down to South America."

JON COPPI AND PETE GOMPERT journeyed into different summer backcountries as an unusually hot and fire-plagued season hung over Utah's northern mountains. Coppi began the summer with a solo mountain climb in Zion National Park along Utah's southern border, a "mission" he chronicled in a blog. "The whole thing was pretty last minute," he wrote, "as I couldn't find a climbing partner for the weekend. I have done some big wall climbing before but this was my first foray into doing it alone."

From the start Coppi found the venture daunting. The only way he could access the climb was on a park shuttle that began running at 7:00 a.m. He needed a permit for staying on a rock wall overnight. By the time he reached the drop-off, it was 8:30 a.m. and the sun was beginning to bake the rock spires. He still had to hike, with all his gear, an hour and a half to the wall he'd decided to scale. "The approach, man that approach sucked," he wrote, displaying some of the same candor he resorted to as test pilot for Black Diamond's JetForce backpacks.

The physical burden was real, Coppi noted, estimating he had an eighty- to ninety-pound pack on his shoulders, including "two 60-meter ropes, a portaledge, a ridiculous amount of rock protection, two gallons of water, food, and sleeping gear filled up my 145 liter haul bag." He wrote: "To be honest, I underestimated how much the approach was going to beat me down." Once he began climbing, with no partner to help manage ropes and anchors, the solo nature of his enterprise was more complicated and time-consuming. After nearly six hours of grueling progress, it began to get dark. "I busted out the headlamp and set up the portaledge at the hanging belay, which is like assembling an erector set while hanging in a harness 400 feet off the ground." By the time Coppi would normally have eaten the chili he brought for dinner, he was too tired to even prepare it, so he "curled into my sleeping bag and called it a day." But sleep proved elusive, as he dangled alone nearly a twelfth of a mile above the earth. "I don't know anyone that actually sleeps well on a portaledge," he observed. "It's kind of this weird state where you're incredibly exhausted but you can't put your mind to rest, so you go in and out of half sleep for the whole night."

When morning came, Coppi decided to pack up and descend. Self-effacing to the end, he judged his adventure "pretty silly" but, like the engineer he is, found useful data in the climb for a future ascent with a partner. "I think I got what I wanted out of it," he concluded, "which was to gain confidence in my systems and experience in the self-reliance of being alone on a wall."

If the Zion climb had emphasized his self-reliance, soon he looked to that trait even more. In July, Coppi decided, after six years, to leave Black Diamond. "I ended up getting another gig with a company called Bare Bones Living," he said over the phone. The job would allow him to do more of the basic ground-up product development work like he'd done with the JetForce backpack. He would miss Black Diamond, but the company "is pretty much all I've known." It was time for Coppi to hone some different skills and pursue new adventures like what he'd done on the Zion wall.

Coppi had already notched a new skiing milestone, climbing and descending Mount Rainier's Emmons Glacier, a 9,500-foot run from the summit to the parking lot. Then it was on to Washington State's Mount Adams to ski the peak's Southwest chutes. "I'd never done any volcano skiing before so that was pretty awesome," he said.

Driving toward Telluride, Colorado, Coppi had his dog, Haydee, a "pure bred mutt" and a dirt bike on board. Between exploring trails in the San Juan Range, he expected to drink a little beer to slake his thirst between future ski adventures. "I've been doing a lot more running," he said. "I'm trying to get into shape for big, long Alpine routes."

If Coppi's summer travels seemed like life's equivalent to bushwhacking through new wilderness, Pete Gompert had to stay focused on a raft of projects at Black Diamond, when summer is high time for putting the finishing touches on products to be rolled out in the fall for winter sales—skis, avalanche beacons, climbing skins, and carabiners. The always-in-demand JetForce backpacks receive more attention as the number of backcountry skiers and boarders grows. "JetForce is the gift that keeps on giving," Gompert said. "It's going to be quite a bit lighter and more compact." A new version of the JetForce—the Tour 26L—was soon to be released, containing a turbine fan powered by "supercapacitors" and offering charging with a micro USB cable and AA batteries.

At Black Diamond, Gompert oversees ski design, so he was putting a lot of time into working with the ski fabricators at the Blizzard Skis factory in Austria, where Black Diamond's skis are machined from Gompert's designs. "I'm kind of the ski guy at Black Diamond and everybody just assumes we take the summer off," he said. "We're always working on something two years out, so there's really no break in the summer." After some previous initiatives at building skis in their own factory in China, Black Diamond settled on the partnership with Blizzard because of the high-quality production. "We send them a CAD file and then two months later a perfect ski shows up," Gompert explained. "The glacier's right there. They're in Mittersill, Austria, so it's kind of like *Sound of Music*. You can drive twenty minutes and ski anything."

It's not as if Gompert deals with one or two ski designs. Black Diamond offers eleven different models with three to five sizes of skis within each vertical so that skiers with varying weight and height can match their needs to a specific ski. Gompert expected to see some prototype skis arrive around the end of July, but that would lead to an ever present dilemma. "We're trying to find a place to ski on 'em. That's always the trick with ski design, you got to get 'em done early, but then you have to sit and wait for snow, or find a glacier somewhere, or go south, which is very expensive."

Gompert had been working with the electrical engineers at PIEPS, the Austrian avalanche beacon company Black Diamond had acquired. As part of the partnership, PIEPS continued to design the electronic "guts" of the beacons, while Gompert and other industrial designers at Black Diamond devised the plastic casings. For the first time, Black Diamond would offer its own branded beacons, the Guide and Recon, while PIEPS would produce two beacons, the Pro and Powder, aimed at bolstering PIEPS popularity in the European market.

When he isn't marshaling product design at Black Diamond, Gompert is on his mountain bike in the snowless ski runs around the Ogden Valley where he lives. He has a small farm there. "We raise a few pigs and cows and chickens and stuff. Work on old tractors. I'm a little more of a redneck than most people at BD, but it's kinda fun." He has two young children. "A lot of guys are obviously climbers at Black Diamond. I've got kids now so I kind of have to choose my hobbies carefully."

When Gompert heard that Coppi would be leaving Black Diamond, he no doubt had an emotional reaction. They had been through a lot with the JetForce pack development, like comrades in a conflict zone. Band of Brothers stuff. "It will be good for him, I think, to get out, get some extra experience," Gompert conceded. "We told him he's always welcome back, so we'll see how long it takes." As if assessing Coppi's test-pilot days, Gompert paused for a moment. "Hopefully, he comes back. I really hope so. I like that guy."

NANCY BOCKINO HAS BEEN ON a seventeen-year summer crusade in Wyoming's Greater Yellowstone ecosystem, near her Jackson Hole home, doing battle with mountain pine beetles. High-elevation trees called white bark pines, *Pinus albicaulis*, or in some places limber pines, *Pinus flexilius*, grow in the mountains of the western United States, Mexico, and Canada. Attributed to climate change, beetles have increasingly decimated the pines, which are considered an essential species in Greater Yellowstone.

Bockino's foe looks disarmingly insignificant, even if it has the ability to ravage a forest in its tenacious migration of malevolence. She describes the beetles as about the size of a grain of rice. An ecologist, in addition to being a rock-climbing and backcountry ski guide and avalanche expert, Bockino has presented at forecaster Craig Gordon's fall conference. In summer, she spends her time scaling pine trees to examine damage and collects pine cones for subsequent replanting. She places substances called pheromones—an attracting scent called verbenone—on the trees to help ward off beetles.

The work isn't sexy, and it's potentially dangerous. The trees she scales range from about thirty feet tall to nearly a hundred feet. "The trees have been around ten thousand years and so have the native beetles," she says, and they are spread across two million acres of Greater Yellowstone. The trees-to-beetles life spans illustrate the angst over the devastation. A white bark pine can live about 1,200 years, assuming it's free of infestation. A mountain pine beetle will exact as much damage as it can in the single year it's alive.

In summer, white pine beetles fly through the forest like microscopic drones targeting living white bark pines. Infesting the trees en masse, female beetles chew through bark and set about laying their eggs. Once the eggs hatch, larvae infest the soft wood all winter, dormant under a tent of bark. When spring arrives, the larvae arise and, using mandibles—ugly snouts that tear at the wood—they feed their way to becoming mature adults. It's like a bug with a raging drug addiction, only the drug is the soft core of the white bark pine. Though the beetles are tiny, the tall trees they attack can succumb to such heavy plundering. Within the beetles' jaws is a fungus—called blue stain fungus—whose fibers choke the trees' cells that produce pitch, the very thing white bark pines generate to repel the beetles.

The work Bockino and others are doing around this problem isn't a case of environmental vanity. "The trees are critical for watershed because they capture snow and it melts more slowly," she says, when it's situated under the trees' lush evergreen branches. "On a big scale, it's about water in the rivers."

There is another factor to the pine beetles that cuts close to Bockino's work with avalanche awareness training. The white bark pine trees help form barriers to avalanche paths; once such trees are killed and removed from those paths, avalanches can gather more speed and affect larger swaths of terrain. Bockino remains optimistic in her combat with the beetles. She has spent so much time clambering among the branches of the pine trees, she attributes a kind of natural instinct to their determination to survive.

"I feel like we're winning because we have this amazing collaboration of land managers and researchers in the Greater Yellowstone ecosystem, and if we can keep enough of them alive they'll know what to do."

CRAIG GORDON HAS MANAGED, FOR the past twenty-eight years, to ski at least one day of every month of every year. As September 2018 approached, that meant

332 unbroken months of making turns on the snow, including runs in July, August, and September, when most snowpack has undergone a slow assassination by the sun.

Imagine it, a deeply suntanned man who, in summer, delights in tank T-shirts, gym shorts, and flip-flops, on a quest for melting snow so he can dial into some ski turns. If most of his life is highly organized—early morning avalanche advisories, overseeing an annual conference of snow scientists, avalanche awareness seminars, television reports—his crusade for summer snow is a little like that of the guy obsessed with hunting treasure on the bottom of the sea. To bystanders, his summer behavior holds an almost mythical dimension: he is like a servant to the deity of snow, always seeking wisdom in the snowpack.

Even Gordon's marriage to Anita holds an annual ritual on the snow. Every May 5 since their wedding in 2001 at Our Lady of The Snows chapel across the street from Alta, the couple suits up in their carefully preserved matrimonial finery and ride the tram up to the summit. In 2019 they celebrated eighteen years by taking pictures, renewing their vows, and leisurely skiing and snowboarding to the bottom.

After his spring backcountry outings, Gordon's pursuit of snow and ski turns in the dog days of late summer is part devotion, part tenacious physical prowess, and part his own private *Guinness Book* record. Sure, he could go easy on himself (though not on his bank account) and fly to New Zealand or Chile, where the snow flies during the American summer. But that would be like cheating. Staying true to the mission of sleuthing snow in his home peaks in the Wasatch means overcoming a certain amount of hardship. It's like religion. Deliverance requires faith for which you pay a price. It's the high noon of skiing commitment, a sacrifice to the snow god's equivalent to a monk's prostrations in the monastery.

"As I explored deeper into the summer skiing mystique, I found few shared my passion of a contiguous season and endless winter," he says about his early years of chasing snow in the summer. "I was working at Brighton year-round and while my fellow coworkers looked forward to boating and river running on the weekends, I was shouldering my skis and kicking steps up lingering snowfields in the Wasatch." Over time, his pursuit of isolated stashes of snow reinforced his commitment every summer. "As I got deeper into stringing seasons together, each summer snowpack had its own personality, and I became a connoisseur of summer snow conditions." In years of deep snowpack, he found unmelted snow in his backyard—Big and Little Cottonwood Canyons. In lean years he had to "trek the dusty hiking trails and endless switchbacks leading to the Mount Timpanogos snowfield."

Gordon's decades-long campaign was not without dangers. "With changing weather and summer thunderstorms, lightning is always a sketchy variable. Several times I've had storms materialize overhead and *bam*, a lightning strike a couple hundred yards away rocks your world." Sometimes it seemed as if his snow dragons had taken refuge in the rock faces looming above where he was skiing. "I've literally been moments away from being at the wrong place and witnessed van-sized slabs of granite peel off of a rock face and bury themselves deep into the summer snowpack. Had I been in the impact zone, I'd instantly become an artifact of geologic history."

In recent times Gordon has noticed his summer hikes into the backcountry are challenged by fewer smears of summer snow. What he does find are small caches, thinner than ever, and higher up the mountain. "I've been able to keep a contiguous string of months together since 1990 without ever leaving Utah, but even in this short stretch of earthly time I can see changes in weather patterns, which are limiting my options."

The marginal winter snows of 2017–18 and ensuing oppressively hot summer presented a particularly difficult obstacle to finding skiable snow. He had managed to get in some turns for July and August, but September stood before him like a daunting, parched stretch of Sahara. "It's almost Labor Day as I ponder leaving the state and heading to the Northwest," he said, "or do I just put an asterisk next to 2017–18? I'm still unsure. What I am sure of is the climate is changing and my past summer skiing will pale in comparison to what is headed our way in the future."

By mid-September he had concluded there was no skiable snow in Utah, the first time in nearly three decades. So he made a quick trek to Oregon's Mount Hood to ski runs on the same snow where Stephan Drake had been testing skis. Gordon's dash to Oregon and back left him facing a last insult from a winter he thought he'd left behind. In his final Utah Avalanche Center posting for the 2017–18 winter, he uploaded photographs of two byways leading into the mountains with signs that read "Road Closed, No Winter Maintenance." But the roads were bare of any snow. "Well this is a heck of a way to run a winter," Gordon captioned the photo. "Picture above . . . a tale of two trailheads. Common theme . . . while still closed, they're both toast and dry for miles."

For Gordon, the winter of 2018–19 was not far off. He had already started thinking about his late October snow conference and preparations for forecasting avalanche dangers from the season's first big snowfall. The winter of 2017–18 had ended with twenty-five avalanche deaths in the United States. Forecasters like Gordon would still be needed to chart the new winter's danger zones. His star continued to rise among his fellow snow professionals. In February 2019 *Backcountry* magazine ran a cover story: "Guardians of The Deep, How 34 Mountain-Safety Icons Have Shaped the Backcountry of Today." As one of the icons, Gordon was cited for his "know before you go" program.

The article featured a photo of Gordon, in a tank shirt and shorts, making a jump-turn above summer snow. The caption read: "Craig Gordon, neither a fan of convention nor shirtsleeves, finds his own groove."

IN THE FIRST FEW MONTHS of my living in Ogden Valley in early 2016, I had become attuned, on most winter mornings, to the sound of avalanche bombings echoing across the valley like sonic booms from an invisible jet fighter. The most crisp detonations came from the south, where Snowbasin ski resort was perched among the nine-thousand-foot peaks that acted like an echo chamber. To the north, just behind my house, was Powder Mountain, where the bombings occurred predictably after storms had passed. But the high buttes muffled those blasts before they spilled over the ridgeline like dying beats from a distant kettle drum.

Over time, I thought of these bombings as the day's evidence of the winter war against Gordon's snow dragons. They lurked up there, waiting for unsuspecting skiers and boarders. And there were more potential victims heading into the backcountry. It may be that, like the rest of terrestrial life on the planet, skiers and boarders in the backcountry, its own white-dusted planet, are finally making their increasing numbers known. This—added to rainfalls higher in the mountains, more dramatic freezing and warming cycles, and unstable snowpack—suggests the propensity of avalanches has reached a tipping point of greater numbers in uncustomary places. Gordon and his colleagues think so. The snowpack, in a sense, is beginning to rebel.

But Gordon, for one, isn't backing down. "There are some days," says his wife Anita, "I just don't know how he keeps going. At home there's always a text, an email, a phone call. It just never ends. What he does would kill a normal human being."

Like Pete Gompert in his Abe Lincoln–Batman T-shirt prowling the halls of Black Diamond; like Stephan Drake in a temple of skis, dreaming of conquering a run on Eduardo's; like Titus Case and his ascetic mornings at Alta, riding the ski lifts alone at sunrise; like Wendy Wagner, a contemporary snow sentry on Alaska's frontier skirting stunning glaciers in the wilderness on a snowmobile; like all the decipherers of the snowpack—Bruce Tremper, Doug Chabot, and many others; like Ian McCammon and Jordy Hendrikx, the adrenaline min-dreaders of all the backcountry human impulses gone awry, Craig Gordon had found his own personal backcountry calling.

Over the many months and three winters I've spent with him, I've come to see that he isn't your average guy grinding out a living in a profession that just happens to be a little different from those New Jersey Joes he went to high school with. No, Gordon has achieved something transcendent. He's ventured out at midnight into the teeth of blizzards, studied snow as it was shot from the atmosphere toward earth, swooped in helicopters over the Uinta mountain peaks attacking avalanches, watched a million flakes a billion flakes settle and link arms, learned to know the crystals' quirks and all their sublime, beautiful moods. Through all this, Gordon has ascended to a higher plane in his understanding of snow—and of people when they venture into snow.

"I started in this for one thing, the love of skiing in the backcountry," he once told me. "But then it grew into such a bigger thing where I realized it is so much bigger than me. . . . It's sort of like Christmas morning when you are giving the gifts back. And the reward is that you know you have contributed a lot to your community."

Yes, Gordon's job description is avalanche forecaster. But as I imagine him making those few desperate if devoted turns through the melting summer snows on Mount Hood, I had come to understand what he really is—a master of ice crystals, an undaunted, indefatigable dragon fighter in the snowy wilderness of winters past and winters still to come.

Acknowledgments

This book could not have been written without the support and guidance of avalanche forecaster Craig Gordon. From the beginning, Craig was "all in" with the project and through emails, phone calls, and personal interactions helped me open doors to people who more swiftly participated in interviews because the introduction came from Craig. Over the three years I worked on the book, Craig and I became friends and I became his ever ready, aspiring ski partner when the rare opportunity presented itself. His skiing mentorship is a crystalline gift to anyone who chances to receive it. As lifelong surfers and skiers, Craig and I realized in our first encounter that we had been living as distant members of the same tribe. Thanks, brother.

I'm indebted to those who read early drafts of the full manuscript and critical chapters and offered sage counsel, technical clarifications, and encouragement: Titus Case, formerly of the Alta ski patrol; Mark Rasmussen, formerly of Snowbird's ski patrol and Petzl mountaineering; Shannon Finch of Park City Powder Cats; backcountry skier extraordinaire Andrew McLean; Doug Chabot, director of the Gallatin National Forest Avalanche Center; Bruce Tremper, former director of the Utah Avalanche Center, and author of definitive books on backcountry skiing, snow science, and techniques of avalanche awareness; Jill Fredston, who along with her husband, Doug Fessler, wrote the internationally acclaimed *Snow Sense: A Guide to Evaluating Snow Avalanche Hazard* and developed early avalanche education methodology that today lives on in worldwide avalanche education; and finally my brother-in-law T.J. Edlich, a non- skier who read the book while flying to and from Spain and declared that "as a weather geek" he was riveted. All these folks' help and kind words kept me forging ahead.

To all the interviewees who gave generously of their time at their homes, in coffee houses, bars, on the snow, and in numerous emails and phone calls, I cannot thank you enough. My hope is that the intent of the book—saving lives in the backcountry—measured up to your votes of confidence in my reporting and writing skills.

To "Team Power" at Mountaineers Books, I could not have been in better hands. My development and copy editor, Amy Smith Bell, handled the manuscript with an atelier sense of craftsmanship rare in today's world. Editor Mary Metz kept all the pieces coming together with a firm hand and helped me fine-tune words that created a whole far greater than the sum of its nascent parts. Jen Grable produced a dynamic cover that fed momentum to the sales and promotions teams who did their work expertly.

From my first email to her with a few sample chapters, Kate Rogers, Mountaineers Books' editor in chief, was a champion for the book. All she wanted to know was when I would be finished. Her skill, confidence, judgment, and counsel are as rarefied as those of the best editors in American publishing.

I'm also grateful for the support of Helen Cherullo, who served as publisher at Mountaineers Books when the book was acquired.

To my sons, Graham and Nick, I am forever grateful that when I said, after thirty-five years, I was leaving my well-paying media job to go back to my first love of writing with nary a paycheck in sight, you guys said: "Do it!"

Finally, to my wife, Marguerite Ulmer, I simply would not be the writer—or more importantly the person—I am without you. If skiers have an affinity for finding the North Star, my life's North Star would not have been as certain and precious had I not found you.

About the Author

Marguerite Ulmer

Edward Power grew up on the southeastern coast of Virginia. He received a B.A. and an M.A. in literature and writing from the University of Virginia and spent a year at Columbia University as a writing fellow in the School of the Arts. He is a former staff writer for *The Philadelphia Inquirer* and *The Virginian-Pilot,* where his assignments took him to the Caribbean, Central America, Asia, and across the United States. Now a full-time freelance writer, he lives with his wife at homes in the mountains of northern Utah and on the Pacific coast of Mexico. Learn more at www.edwardpowerwriter.com.

recreation • lifestyle • conservation

MOUNTAINEERS BOOKS is a leading publisher of mountaineering literature and guides—including our flagship title, *Mountaineering: The Freedom of the Hills*—as well as adventure narratives, natural history, and general outdoor recreation. Through our two imprints, Skipstone and Braided River, we also publish titles on sustainability and conservation. We are committed to supporting the environmental and educational goals of our organization by providing expert information on human-powered adventure, sustainable practices at home and on the trail, and preservation of wilderness.

The Mountaineers, founded in 1906, is a 501(c)(3) nonprofit outdoor recreation and conservation organization whose mission is to enrich lives and communities by helping people "explore, conserve, learn about, and enjoy the lands and waters of the Pacific Northwest and beyond." One of the largest such organizations in the United States, it sponsors classes and year-round outdoor activities throughout the Pacific Northwest, including climbing, hiking, backcountry skiing, snowshoeing, camping, kayaking, sailing, and more. The Mountaineers also supports its mission through its publishing division, Mountaineers Books, and promotes environmental education and citizen engagement. For more information, visit The Mountaineers Program Center, 7700 Sand Point Way NE, Seattle, WA 98115-3996; phone 206-521-6001; www.mountaineers.org; or email info@mountaineers.org.

Our publications are made possible through the generosity of donors and through sales of 700 titles on outdoor recreation, sustainable lifestyle, and conservation. To donate, purchase books, or learn more, visit us online:

MOUNTAINEERS BOOKS
1001 SW Klickitat Way, Suite 201 • Seattle, WA 98134
800-553-4453 • mbooks@mountaineersbooks.org • www.mountaineersbooks.org

An independent nonprofit publisher since 1960

OTHER TITLES YOU MIGHT ENJOY FROM MOUNTAINEERS BOOKS

TRACKING THE WILD COOMBA
Robert Cocuzzo
"Doug Coombs was an inspiration to me and so many others on and off the mountain. Now, here is an insightful look at the life of a legend."
—Jimmy Chin

STAYING ALIVE IN AVALANCHE TERRAIN, 3RD EDITION
Bruce Tremper
"No one who plays in mountain snow should leave home without having studied this book." —*Rocky Mountain News*

CROSSING DENALI
Mike Fenner
An inspirational tale of a novice climber who attempts the tallest mountain in North America during a particularly stormy season.

FOUND
Bree Loewen
"A truly inspirational book about the incredible people who risk their lives to save others." —*Bustle*

OVER THE EDGE
Greg Child
"Riveting and meticulously researched, *Over the Edge* recounts a climbing expedition gone horribly wrong." —Jon Krakauer

ARCTIC SOLITAIRE
Paul Souders
"Souders immerses his audience in a captivating adventure, offering an intimate view of both the perils of lone exploration and the stunning beauty of the wild. The result is a rare, enchanting treat for any reader." —*Foreword Reviews*

www.mountaineersbooks.org